Gender Violence in Ecofeminist Perspective

This book aims to begin an eco-centered, eco-feminist informed discussion about the ways in which our relationship to "nature" is bound up with gender, patriarchy, and violence.

Ecofeminist scholars study the interconnections between gendered relationships of domination among humans, between humans, and between humans, nonhumans, and the earth. It is in this ideological and structural tangle between humans and the environment that a deeper understanding of gender violence is possible. Ecofeminism offers analytical possibilities for understanding a "logic of domination" which sustain a whole host of problems, including the interrelated oppressions of gender violence and exploitation of the more-than-human-life world. In this book, Gwen Hunnicutt brings into dialog ecofeminism and gender violence. Ideological components, such as speciesism and the belief that the earth and its nonhuman inhabitants are ours to exploit, inform a host of other social practices, including interpersonal violence.

A portion of this book is devoted to exploring the ways in which patriarchy is foregrounded by another hierarchy—human domination over "nature". Thus, gender violence stems from a logic of domination that is built on the domination of nature and the domination of the Other "as nature". As this blueprint of oppression repeats itself where there are vectors of difference, the chapters ultimately connect these oppressions by showing the inextricable bind of violence against humans and the more-than-human-life world. This book will serve as a resource for scholars, activists, and students in sociology, gender violence and interdisciplinary violence studies, critical animal studies, environmental studies, and feminist and ecofeminist studies.

Gwen Hunnicutt is an Associate Professor of Sociology at the University of North Carolina Greensboro. Professor Hunnicutt received her PhD from the University of New Mexico and studies various dimensions of gender violence. Her writings have appeared in various journals, including *Sexuality and Culture*, *Violence Against Women*, *Gender Issues*, *Journal of Family Violence*, and *Journal of International Women's Studies*.

Routledge Research in Gender and Society

For more information about this series, please visit: https://www.routledge.com/sociology/series/SE0271

Gender Violence in Ecofeminist Perspective

Intersections of Animal Oppression, Patriarchy and Domination of the Earth

Gwen Hunnicutt

Routledge
Taylor & Francis Group

LONDON AND NEW YORK

First published 2020 by Routledge

2 Park Square, Milton Park, Abingdon, Oxon, OX14 4RN
605 Third Avenue, New York, NY 10017

Routledge is an imprint of the Taylor & Francis Group, an informa business

First issued in paperback 2020

British Library Cataloguing-in-Publication Data
A catalogue record for this book is available from the British Library

Library of Congress Cataloging-in-Publication Data
Names: Hunnicutt, Gwen, author.
Title: Gender violence in ecofeminist perspective : intersections of animal oppression, patriarchy and domination of the earth / Gwen Hunnicutt.
Description: Abingdon, Oxon ; New York, NY : Routledge, 2020. | Series: Routledge research in gender and society | Includes bibliographical references and index.
Identifiers: LCCN 2019030473 (print) | LCCN 2019030474 (ebook)
Subjects: LCSH: Ecofeminism. | Violence. | Speciesism. | Patriarchy. | Human ecology.
Classification: LCC HQ1194 .H86 2020 (print) | LCC HQ1194 (ebook) | DDC 304.2082—dc23
LC record available at https://lccn.loc.gov/2019030473
LC ebook record available at https://lccn.loc.gov/2019030474

ISBN: 978-1-138-49384-1 (hbk)
ISBN: 978-0-367-77691-6 (pbk)

Typeset in Times New Roman
by codeMantra

To all nonhuman animals in captivity

Contents

Preface

Because environmental crises related to climate change will eventually impact every area of social life, all scholars will need to heed environmental concerns. The other-than-human-life world is often rendered invisible in descriptions and explanations of social life. In the age of the Anthropocene, scholarship of all kinds will need to illuminate the connectedness between human behavior and our relationship to the ambient natural environment. In the next 50 years, we will see the world's population increase beyond nine billion, watch green spaces and species disappear, and witness large-scale human suffering. What does it mean to be a scholar—in any field—at the dawn of global ecological crisis? I endeavor here to reach many audiences: sociologists, gender violence researchers, environmentalists, feminists, ecofeminists, and the non-academic public. This book is also written for anyone who wants to heal the earth and heal our bodies, and to protect the earth, nonhuman animals, and people from violence. This book is for those who are open to the idea that violent behavior, which always manifests in gender-specific ways, is an outgrowth of our relationship with the earth/ nature/environment.

In these pages, I attempt to open an eco-centered, eco-feminist informed exploration of connections between humans' relationship to "nature", gender, patriarchy, and violence. Ecofeminist scholars study the interconnections between gendered relationships of domination among humans, between humans, and between humans, nonhumans, and the earth. It is in this ideological and structural tangle between humans and the environment that a deeper understanding of gender violence is possible. Ecofeminism offers analytical possibilities for understanding a "logic of domination" which sustain a whole host of problems, including the interrelated oppressions of gender violence and exploitation of the more-than-human-life world. My goal with this project is to bring into dialog ecofeminism and gender violence, which takes as a starting point ontological dualisms of devaluing the "Other" and a contingency between a gendered environment and gendered bodies. Ideological components, such as speciesism and the belief that the earth and its nonhuman inhabitants are ours to exploit, inform a host of other social practices, including interpersonal violence.

While patriarchy is widely theorized in the gender violence literature, a major portion of this book is devoted to exploring the ways in which patriarchy is foregrounded by another hierarchy—human domination over "nature". Thus, gender violence stems from a logic of domination that is built on the domination of nature and the domination of the Other "as nature". As this blueprint of oppression repeats itself where there are vectors of difference, I ultimately connect these oppressions by showing the inextricable bind of violence against human and nonhuman animals and environmental degradation. I conclude that the prevention of gender violence occupies the same ethical ground as our obligation to the more-than-human-life world. I argue for a posthumanist, feminist intervention in the sociological study of violence. I conclude that our work on a nonviolent future is inseparable from our progress on a green future and that there can be no social justice without ecological justice.

I am committed to a practice of nonviolence in all areas of my life: in communication, consumption and how I move about social space. I strive (imperfectly) to bring this ethic to all my relationships, both personally and professionally. Nonviolence is not just something to teach students in a classroom, but also a path of integrity and an ethic for life. The main impetus for studying gender violence throughout my career has been rooted in a belief that all humans and nonhumans have the right to safety, to live free from bodily harm. As a privileged, white, middle-class female, I enjoy the ability to feel safe in the world. I am continuously aware that my experience of freedom of movement and feelings of safety are very much connected to my resources and advantages which are bound up with my race and nationality. It is my hope that this work contributes in some small way toward a collective aspiration for nonviolence and safety for all creatures, human and nonhuman.

Acknowledgements

I would like to express my gratitude to all the people and institutions that have made this book possible. I am indebted to both the animal emancipation movement and anti-violence movement and individual scholars who are advocates for nonviolence and defenders of peace. The University of North Carolina Greensboro generously awarded me with research leave for the Fall 2018 semester, which provided time to work on this manuscript. I am deeply appreciative for this beneficial gift. My Department Head Dave Kauzlarich was endlessly supportive and believed in this project from start to finish. My colleague and friend, Steve Kroll Smith, provided helpful advice, support, and encouragement. Deepest thanks to Daniel Rhodes, who first introduced me to the Ecofeminist literature over ten years ago and has since been a close collaborator and dear friend. My son, Alex Hunnicutt-La, who is the source of my inspiration to want to make the world a better place. My dog, Neo, who loyally held vigil beside me during the writing of this book, patiently witnessing the construction of this entire manuscript. Finally, my deepest love and appreciation to my husband, Keith Buckner, who was supportive, generous, patient, and understanding during all the hours that I was away working.

Chapter 1

Introduction

Ecofeminism, "nature", and gender violence

Ecofeminism has been gaining theoretical ground across disciplines as diverse as public health, philosophy, ecology, and various sciences (Eaton & Lorentzen, 2003; Seager, 2003). Ecofeminism operates with the idea that the injustices directed at marginal social groups share similar ideologies with those that legitimate the exploitation and degradation of the environment (Adams & Gruen, 2014; Thompson & MacGregor, 2017). Ecofeminism, also called feminist environmentalism, works to expose those gendered assumptions, performances, and practices that undergird human relationships with the more-than-human-life world (Gaard, 2017). Ecofeminists are taking on issues that range from global political economic issues, to animal rights, to environmental racism, and the environmental origins of illness (Alaimo, 2000; Harper, 2011; Rocheleau, Thomas-Slayter, & Wangari, 1996). Ecofeminism is, therefore, a multi-issue, anti-racist, inclusive movement with a diverse constituency (Gaard, 2017; Mies & Shiva, 2014). Even though environmental feminisms have the potential to generate insights and recommendations for many areas of social life, scholars have yet to fully tap its analytical potential to deepen our understanding of gender violence. This book aims to begin an eco-centered, eco-feminist informed discussion about the ways in which our relationship to the environment is bound up with gender, patriarchy, and violence.

Environmental feminist scholars study the gendered intersections of various relationships of domination those among and between humans, nonhumans, and the earth (Merry, 2009; Mies & Shiva, 2014; Plumwood, 1993). The ideological and structural conditions between humans and the environment offer insight into gender violence. This insight is achieved by exposing the logic of domination that sustains a whole host of problems, which includes the interrelated oppressions of gender violence and degradation of the more-than-human-life world. As ecofeminism and gender violence are brought into dialog, we begin to see how a gendered environment and gendered bodies are mutually constituted. Drawing on three key literatures—Gender Violence (as well as the overlapping Violence Against Women literature), Green Criminology, and

Ecofeminism or Environmental Feminism—this book is an effort to encourage cross fertilization between these fields which strives to tease out links between gender violence, our relationship to the non human-life world, patriarchy, and the domination of our planet. Ultimately, this text rests on the assumption that human behavior is conditioned by our relationship with the non-human-life world; a relationship that is very gendered. This book, therefore, excavates the intersections of multiple patriarchies, ecology, and gender violence.

Gender violence is on the ecofeminist agenda, and gender violence may potentially be challenged through a strategic alliance between environmentalism and feminism. Indeed, ecofeminists have revealed those multiple linkages within the devalued category of the "Other" (Plumwood, 1993). Ecofeminists have pointed out how the association of characteristics between oppressed groups functions to reinforce their subjugation. For example, the association between women, or any "Othered" group, and animals emphasizes their joint inferiority as they are simultaneously being naturalized, feminized, and animalized (Adams, 1990, 1993).

While this text makes extensive use of ecofeminist perspectives, fundamentally, this book constitutes a sociological analysis of gender violence, exploring how this phenomenon has roots that extend into human domination of nature. Ideological components, such as speciesism and the belief that the earth and its nonhuman inhabitants are ours to exploit, inform a host of other social practices, including interpersonal violence (all of which, I will later argue, are gendered). Speciesism is an ideology that regards one species (humans) as superior and other species (nonhuman) as inferior. This belief system legitimates prejudice and discrimination. Speciesism is lodged within a larger hierarchical apparatus where entire populations are deemed inferior based on stereotypical assumptions about their intellect and physicality. Speciesism involves the position that animals—"a huge and unwieldy category that encompasses creatures as diverse as mosquitoes, jellyfish, dogs and orcas—are understood to be unquestionably inferior creatures" (Taylor, 2017, p. 19). Species is a constructed system of power which intersects with other hierarchies defined by gender. Understanding these intersecting systems of power can reveal how violence is threaded through the practice of othering animals, women, people of color, and other subordinated groups.

The inferiority articulated by speciesism is constituted not just through institutional processes where animals are incarcerated, oppressed, and exploited (i.e. research labs, slaughterhouses, hunting, factory farming, zoos) but also discursively and materially constituted in cultural processes. The oppression of animals—legitimated by species relations of "superior and inferior" and the institutionalization of animal subjugation—is now a normative practice, and the maintenance of such a social system of human domination requires violence (Cudworth, 2015). This understanding of species as relational and institutionalized helps us to think sociologically about a multitude of associations between human violence and oppression of domesticated animals. Combating gender violence, as well as other interlocking oppressions, requires structural and ideological changes in our oppressive relationship with nature. Challenging those power arrangements that inform gender violence also means challenging power relations with the biosphere.

It is no coincidence that we frequently conflate metaphors of human and nonhuman violence, such as "harvest of bodies", "beating a dead horse", "I have a bone to pick with you", and "kill two birds with one stone". In this same discursive thread, we often see violent domination of nature rationalized by feminizing the natural world in such phrases as "rape of the earth" and "mother nature is conquered". These metaphors unmask a system of relational power where ecological and gender dominations are co-constituted with other kinds of complex forms of domination, and where gender violence is an outgrowth of our relations of domination with the earth. Just as gender violence stems from the same system of domination of the earth, ameliorating the problem of gender violence is yoked to advancement of eco-justice. There is no solution to gender violence and the ecological crisis if our fundamental model of relationship continues to be one of domination, precisely because patriarchal systems are bound up with other hierarchical systems. As such, an understanding of gender violence must be situated within larger fields of hierarchy where humans dominate not just each other, but also nature.

This work ultimately concludes that there is no social justice without climate and environmental justice. For all animals—human and nonhuman—to survive and thrive and live fully, our biosphere needs to be healthy. These struggles for nonviolence, for the human, for the animals, and for the environment, are interwoven. To end the exploitation of nature requires a dismantling of power structures that also undergird human exploitation. Since systems of oppression are often mutually reinforcing, ecofeminism is useful as it works at the roots of multiple dominations.

Notes on terminology

Both terms "gender" and "environment" are value-laden and take on different meanings in varying contexts. Both terms are products of particular power relations and are therefore understood differently over time and place. The term **environment** typically indicates the nonhuman natural world, positioned in such a way as to make a distinction from human societies (MacGregor, 2017). Synonyms for the environment include ENV, nature, biosphere, ecosystem, ecosphere, and planet earth. While the environment is frequently offered as a binary construction (i.e. environment and society), I aim to highlight the interconnectedness of human- and nonhuman-life worlds. Thus, the "environment" here is a more inclusive term, where humans are embedded in a more-than-human natural ecosphere.

The term **gender** refers to a social construct that is distinguished from biological sex. Gender is a social category and concept which organizes social life. All humans are gendered, sexual beings. Our life experiences and interactions are informed by our gendered expression. Gender is a social status that is infused with power relations, particularly as it intersects with other markers of difference. Gender can be a space of liberation or a construct that is used to oppress and marginalize. Gender can be an identity that an individual rejects, embraces or struggles with. Gender is therefore not just a personal identity, but also a lived experience and an ideology that is political in the sense that it shapes our cultures, our histories, and our system of rewards and privileges. Gender is used to reinforce categories of difference, sometimes resulting in pathologizing and infantilizing women, burdening men with impossible expectations for achievement of masculinity, punishing transgender or queer individuals for transgressing normative gender boundaries, or marking women as vulnerable and dependent. In other words, gender is a social force that is pervasive in how it has shaped our identities, privileges, risks, rewards, political power, opportunities, experiences, reactions, histories, politics, and the modern world (Bradley, 2013).

When I conceptualize gender as "women" or "men", these references should not be interpreted as denying plurality or change. I am not referring to a singular or static condition. When scholars deploy the term "woman", they may intend for the term to include all females, but instead often represents only a narrow group—the most privileged, white, heterosexual, able-bodied, middle-class women. Gender is not fixed, but it instead varies across time, location, culture, race, class, and ability. However, when deployed in a binary manner, gender posits an opposition. The lived experience of masculinity will be different for a heterosexual, Latino man living in an urban U.S. city than a Muslim-American man living in the rural Midwest, so it is important to respect differences when studying gender. Even as gender expressions, identities, and performances exist as pluralities, gender ideology (common beliefs about gender) operates as a diffuse worldview that influences the operations of institutions and power relations. Gender does not trump other identities, such as race and class, but instead, our markers of difference are co-constructive. They are mutually constituted where, for example, gender forms meanings about race and race forms meanings about gender. Indeed, much of this work in feminist environmental studies "problematizes the concepts of 'women' and 'nature' and explores how gender (as a social category and power relation) shapes and is shaped by inter-human relations as well as human relations with other species and environments" (MacGregor, 2017).

The conceptualization of gender violence warrants a lengthy discussion. I elaborate on this topic in the next section. In this text, I use **gender violence** as a way of seeing all violence through the prism of gender. This approach is not so much about gender differences among the actors in violent scenarios and more about gendered social processes, gendered social contexts and the larger gender order where violence occurs. In other words, this approach is less about individual identity and more about how violence is shaped by gendered hierarchies, structures, situations, and beliefs. In the same way that all our life events are gendered in the ways that gendered situations, structures, and ideologies impact our actions and experiences, so too is violence (Throsby & Alexander, 2008). Therefore, all violence is gendered. When referring to violence, I am referencing bodily harm. Violence is a term used widely to evoke a whole host of harms and injustices. We sometimes talk about poverty being violence or even words as violent. In this text, I am using violence to reference a corporeal violation—physical harms against the body, including homicide, sexual assault, rape, and physical abuse. Violence, then, as it is used here, is centrally concerned with bodies (Aldama, 2003).

Typically, when the term **animal** is used, it references nonhuman animals. Humans, of course, are animals too. The term "animal" is typically employed in a dualistic manner where the human/animal binary is infused with insult and inferiority conferred on animals as "below" humans. Given this derogatory and charged use of the term "animal" I choose to use the terms, human–animal and nonhuman animal at various times throughout this text with the aim of disrupting this dualism, highlighting our biological similarities and reclaiming humans' animality.

In this text, I am using the term **ecofeminism** to describe the scholarship, writing, activism, and ways of living that hold important the connections between human and nonhuman animals as well as the gendered environment (Adams & Gruen, 2014; Ashcroft, 2013). Feminist analysis of environmental issues may be called ecofeminism, gender and environment, environmental feminism, or feminist environmental justice (Sturgeon, 2017), all of which are used interchangeably in this text.

The term **state** is used throughout this text and refers to an elite governing body that includes agenda setters, leaders, and powerful institutions. This apparatus includes criminal justice, American foreign policy, and political campaigns. The state is a dynamic institution, not a monolithic actor with a fixed set of intentions (Watson, 1991), yet the state possesses governmental power to enforce and enact laws to maintain an economic system that represents their interests (Gramsci, 1992). The state is also highly influential among civil society: institutions such as religion and education. Indeed, the state is a purveyor of particular ideologies which are disseminated to serve its interests (Sassoon, 1980). Under such an arrangement, civil society institutions may operate as tools of the state which function to advance an ideology aligned with state interests (Nieto-Galan, 2011).

In this text, I define **capitalism** as a global political-economic system characterized by corporate and state actors who share the goal of endless capital accumulation and are acting in service of integrated transnational markets (Kotz & McDonough, 2010). Cultural and political convergences across countries are key to both the configuration and reproduction of the hegemonic global capitalist market. **Neoliberal ideology** is a political belief system deployed to sustain and reproduce this global economic apparatus (Harvey, 2005). This neoliberal political economy depends on the freedom to exploit the environment (Steger & Roy, 2010). In conceptualizing **hierarchy**, I rely on Murray Bookchin (2005, p. 68) who sees hierarchy as a

> complex system of command and obedience in which elites enjoy varying degrees of control over their subordinates without necessarily exploiting them. Such elites may completely lack any form of material wealth; they may even be dispossessed of it, much as Plato's "guardian" elite was socially powerful but materially poor.

Bookchin (2005) adds further nuance to the condition of hierarchy as not just a social condition but also a sensibility or a state of consciousness and that this frame of reference can exist in both personal exchanges and social experiences. While exploitation is not always a feature of hierarchy, the conditions that hierarchies create make exploitation possible.

Conceptualizing gender violence

Violence between human animals is a destructive anthropocentric practice. Historically, gendered patterns in violent behavior were (and still are among some scholars) named, violence against women. This naming of "violence against women" was a very important achievement of the second wave of feminist politics. Because violence against women was historically invisible, ignored, trivialized as unimportant and dismissed as a "private matter", this naming of the problem resulting in the issue finally being acknowledged by the state, medical, and criminal justice practitioners. More recent work, however, has begun to problematize the phrase "violence against women". As much as the phrase "violence against women" was necessary and useful for decades, scholarship has advanced in such a way that demands more inclusive language that strives to avoid problematic binaries of exclusion.

The effort to avoid either/or dichotomies of man/woman, agency/vulnerability, perpetrator/victim includes renaming "violence against women" to "gender violence". There are several additional reasons why the phrase "violence against women" might be problematic. "Violence against women" suggests a set of power relations which "assume male strength and female weakness" (Barnes, 2008, p. 38). This phrase also implies that the unnamed, but assumed male are universally violent. The representation of "violence against women" also functions to afford agency to men while denying it to women. Where women are not actors in violent scenarios, violence becomes the exclusive province of masculinity (Howe, 2008; Shepherd, 2008).

Another potential problem with the phrase "violence against women" is that certain types of violence against women (especially sex trafficking) have become hypervisible and thus have become spectacle. The politics of visibility are in play once again, but this time the issue is hypervisibility (Nordstrom, 1999). The irony is that the more visible violence against women becomes, the more likely the origin of such violence is obscured. The violence that has been turned into spectacle becomes a distorted source of fascination and urgency. Hypervisible scenarios feed into urgent interventions, which often end up ignoring systemic violences that support our lifestyle, in particular "physical violence and coercion that sustains relations of domination and exploitation" (Žižek, 2008). In other words, as violence against women becomes a sensational subject of public fascination, we become distracted from the structural causes of the interpersonal violence. Žižek cautions us to move away from the seductive calls to stop violence—the "humanitarian" urgent agenda, and instead learn about the "complex interaction of all three modes of violence: subjective, objective, and symbolic". Renaming violence against women to "gender violence" is another mode of resisting the fascination of subjective violence.

Sensationalizing violence happens largely because of a feminized story line, where there is a helpless, forlorn victim in need of rescue. This is consequential because not only might we be feeding an appetite for spectacle, but we might be unwittingly contributing to a "fake sense of urgency" (Žižek, 2008). The term "tyranny of urgency" is seen in feminist international relations (IR) theory (Sjoberg, 2010). The idea is that long-term perspectives are eclipsed by the pressure of immediate demands (Enloe, 2004; Jacobson, 2013). As an ameliorative technique, we can situate the study of gender violence within a larger landscape where environmental, political and economic conditions that create violence become visible. When seeking to uncover oppression and human suffering, a broader context is important to acknowledge (Abu-Lughod, 2013; True, 2012).

The conceptualization of "violence against women" simultaneously renders violence against men invisible and homogenizes all men as brutal. The term naturalizes the link between men and aggression, and it presents a rather crude depiction of men as uncontrollably aggressive (Everhart & Hunnicutt, 2013). By focusing on women as victims, but not men, war is still a justifiable endeavor and men are also not extended the right to have violence-free lives, nor are they offered the same moral outrage given to female victims. It is still the case that much scholarship continues to focus on violence against women, which results in the reproduction of binary structures of female victims and male perpetrators, while the scope and impact on male and other-gender victims are overlooked. Consider the work by Chris Dolan whose 2014 study of Congolese refugee men in Uganda showed that 38.5% experienced assaults at some point in their lives and 13.4% of male refugees reported at least one experience of sexual violence in the preceding year. These kinds of statistics are rare in the literature. Part of that stems from the "feminization of violent victimization". The dominant narrative of violent victimization is closely bound up with "women only".

Additionally, the phrase "violence against women" imbeds another hidden hierarchy—one of vulnerability, rescue, retaliation, and protection. Where language evokes chivalry or protecting women, then violence is required to maintain that hierarchy of protection (Chavetz & Dworkin, 1987). In this rescue scenario, retaliatory violence is expected. In this sense, the prevention of violence (against women) perpetuates other forms of violence (against "perpetrators"). Indeed, rates of domestic violence are higher during wartime, and military personnel tend to have higher violation rates than the general population (Nordstrom, 1999).

In order to engineer more complexity into our understandings of gender violence, the focus should shift toward those gendered social processes that shape crime, rather than a narrative that casts all men as violent (Shepherd, 2008). Indeed, violence *is* gendered. Gendered norms and practices inform the context in which violence occurs. Even material objects that play a role in violent events are gendered. For example, guns are gendered. Guns are associated with penises: "they point, they ejaculate, they penetrate, they can be shocking when exposed and they usually can be found adorning men (around the waistline)" (Homsher, 2001, p. 31). Of course, gender is not the only dimension of social life that impacts our experience with violence, which is why an intersectional approach should be taken up as an indispensable analytical tool (Crenshaw, 1991). Popular narratives of gender violence, however, fail to account for the roles that class, race, nation and power play in violent events. Historically, scholars and activists who worked on the eradication of violence against women often neglected an intersectional approach. In these narrow configurations of women as victims, race, sexuality, and class were often ignored, while whiteness and privilege frequently went unchallenged (Hooks, 1981).

The separation of public and private spheres of action for men and women is reflected in our study of gender violence. Efforts to counter the "dominant androcentric male-as-norm" paradigms can be incorporated into our study of gender violence by challenging the public/private divide seen in our association of women with "domestic" violence and men with violence in the public sphere. These associations are implicit in our use of the term "violence against women". The phrase conjures up a domestic violence scenario, a private place. Meanwhile, the violence that men exact in public spheres goes by different names: terror, assault, homicide. The term "gender violence" has the potential to collapse these embedded gender stereotypes and begins to take up all violence for serious feminist study, where gender analysis becomes a lens for interpreting and understanding all types of violence.

This work calls for both a differentiated understanding of gender violence and to see ecological degradation and domination as a gendered process.

Gender violence is a serious issue and an important subject of study. We must be careful not to diminish the seriousness of the problem or lose sight of the achievements of anti-violence women's movements who, over the last 40 years, have been wildly successful in raising awareness among criminal justice practitioners regarding issues of violence against women. It is equally important not to dilute our understanding of gender violence by getting endlessly caught up in critique. But because the topic of gender violence carries such an enormous potential for distortion and appropriation, theorists could work to avoid the pitfalls, which often include a heavy focus on the body. Because violence is corporeal, involving harm to the body, the political context in which violence occurs is often obscured by focusing too heavily on physical realities. The very phrase "violence against women" conjures up the corporeal condition of being a woman—sexed bodies that are often constructed as defenseless, exploitable, and ready for attack (Aldama, 2003). The narrow focus on the body tends to obscure those gendered social processes that contribute to violent actions. Too much focus on the body may lead us back to essentializing, naturalizing territory.

When scholars make a distinction between men and women, gender is not studied in relation to each other, but separately. If we recount only the different correlates, experiences and impacts of men and women, we miss what happens to gender relations, we miss the ways in which violence affects and is affected by power relations between genders. Having described the tension among the use of these various terms, and while the phrase "gender violence" will be employed in this book, I review the literature on violence against women and employ the language of the scholars whose work I am referencing. While many scholars have moved away from the phrase "violence against women" and adopted the more inclusive term, "gender violence", there is not an agreed upon definition of this somewhat nebulous concept. And so, at this time, both terms are widely used in the literature. As I reference the literature, I will be drawing on conceptual intersections of work that employs the terms "violence against women", "men's violence", "gendered or gender-based violence", and "sexual violence", seeking to forge understandings of violence across these conceptualizations.

A study of gender violence, then, considers how gender (as norms, beliefs, social processes, and practices) shape violence. Acts of violence are gendered, but so are responses to gender violence, as well as the ways in which gender violence comes to be understood. Since the response to gender violence can perpetuate the very problem we are trying to solve, it is crucial to explore how gender is embedded in both violent acts and the response to violence. Finally, because the individuals who engage in violence are usually (although not always) themselves in a socially marginal position, it is important that analysis proceed in ways that do not pathologize people or entire societies. It is more likely that the least powerful men engage in violence precisely because they are caught in a system where they face social pressure to accrue more power, where their own disenfranchisement leads them to seek alternative ways to redeem their "wounded masculinity". Moreover, in our quest for gender equality, we run the risk of imposing particular values and expectations on others where gender equality is used as a tool of cultural domination. Ultimately, a study of gender violence necessitates that we are careful that our aspirations toward a more just and equitable world are not deployed in such a way that certain cultures or individuals are regarded as "backward".

What is ecofeminism?

Environmental philosophy is a relatively new discipline, appearing in the 1970s as environmental problems were particularly urgent (Gottlieb, 2005). Environmental philosophers identified anthropocentrism as one of the main sources of environmental degradation (Pellow & Brulle, 2005). Anthropocentrism is the prejudicial favoring of humanity and a devaluing of the environment, evident in the hierarchies we form that elevate the human (e.g. dog and his "master"). Moreover, an anthropocentric attitude values the other-than-human-life world only to the extent that it has utility for humans (e.g. resource extraction). A pivotal text in the development of environmental thought, and later Western Ecofeminism, was Rachel Carson's (1962) *Silent Spring*. Rachel Carson's book (1962) was the first to unearth links between militarism, science and environmental destruction and the subsequent costs to human health. Rachel Carson (1962) was one of the first to point out the environmental calamity of our attempts to control nature through the indiscriminate use of pesticides. Carson (1962) exposed the myth of scientific and human "progress" and critiqued the lack of humility among humans who see themselves as removed from nature.

Ecofeminism in the USA is an outgrowth of the mainstream environmental movement and grassroots politics (Griffin, 1978; Warren, 1996). While mainstream environmental philosophers see anthropocentrism as the key source of ecological degradation, early ecofeminists blamed androcentrism, or domination by men, as the root cause of this environmental crisis (Merchant, 1994; Plant, 1989). Ecofeminists are both scholars and activists who have been involved in both feminism and environmental (toxic waste, deforestation) lobbying efforts (Adamson, Evans, & Stein, 2002; Turpin & Lorentzen, 1996). Ecofeminism became even more visible after nonviolent direct-action movements against nuclear power and weapons (Kraus, 1993; Zimmerman, 1994). The movement really began to take off after the Women and Life on Earth: Ecofeminism conference in 1980. Influenced by the writings of Susan Griffin, Charlene Spretnak (1982), Ynestra King (1989), and Starhawk (1979), a set of political positions were created, and the term "ecofeminist" began to take root. Next, the Ecofeminist Perspectives: Culture, Nature, Theory Conference in March 1987 started the rapid growth of ecofeminist art, political action and theory that has been carried on to this day (Adams, 1994; Haraway, 1989; Merchant, 1980; Plumwood, 1993; Walker, 1988). At these early stages, the main intellectual project of ecofeminism was to interpret the connections between the unequal status of women and nature. This key relationship—the joint devaluation of women and nature—is also the key factor that resulted in ecofeminism being criticized, a subject I address in the next section (Epstein, 2002; Sturgeon, 1997).

Following the pivotal 1987 conference, ecofeminist theory and political action proliferated (Sturgeon, 1997). The term "ecofeminism" was now in play and the movement was recognized as an effort to merge feminism with environmentalism, but also nontraditional spiritualities, anti-imperialism, antimilitarism, antiracism, and animal rights (Sturgeon, 1997). While Ecofeminism, or Environmental Feminism, has sought to explore the links between ecological destruction and gendered injustice by claiming that the domination of nature and the domination of women are structurally linked (Thompson & MacGregor, 2017), ecofeminism does more than that. Ecofeminism also questions the tendency to approach culture as an exclusively human terrain, calling attention to discourse that tends to essentialize "culture" as exclusively human (Twine, 2014). Additionally, ecofeminism has a solid ethical tradition that promotes nonviolent practices by drawing attention to intersecting structures of power that promote speciesism, such as colonialism, ableism, racism, sexism, and heteronormativity (Adams & Guren, 2014). Indeed, "an intersectional analysis of capitalism, rationalist science, colonialism, racism, (hetero)sexism, and speciesism has always been central to feminist environmental scholarship" (MacGregor, 2017). While socialist feminists are credited with first exposing the interconnections of sexism, racism, and classism (Jagger, 1983), ecofeminist thought later brought attention to the additional oppressive structures of speciesism, patriarchy, and harm against the natural world (Zimmerman, 1994).

While ecofeminism has made broad theoretical interventions in a range of co-constituted oppressions, at its inception ecofeminism was most instrumental in exposing the link between the domination of women and the domination of nature. It is no coincidence that the earth and nature are personified as female, and images of nature are often used as metaphors for femininity (Warren, 2000). This objectification of the earth casts nature as passive. If women are equated with nature, their struggle for emancipation is bound up with our ideas about nature as disembodied. Indeed, Peggy Reeves Sanday (1981a, 1981b) found that rape occurred very rarely in societies where nature was held sacred, signaling that respect for and value of women is yoked to our reverence for nature. One key tenant of ecofeminism is that, historically, women have been denied full subjectivity partially because of their symbolic connections to the earth and their "alleged animalistic nature", notions largely rooted in Christian theology (Gudmarsdottir, 2010). As Salleh (1997) observes, "the basic premise of ecofeminist political analysis is that ecological crisis is the inevitable effect of a Eurocentric capitalist patriarchal culture built on the domination of nature and the domination of Women 'as nature'" (p. 12). There is currency in equating women with nature—it legitimates their domination. Of course, women aren't closer to nature in any ontological sense, but rather, our current social configuration dictates that the accrual of power involves a "distancing" from nature. More mature writings in ecofeminism focus not just on women, but an array of intra-human differences (race, class, age) and their links to a devalued environment.

Ecofeminists also point to the colonization of the more-than-human-life world'by means of reason. Luke Martell (1994), in *Ecology and Society*, examines the "culture of control" characterized by the Enlightenment. Enlightenment thinkers, most notably Francis Bacon, saw the "taming of nature" as serving the betterment of humanity, manipulating an "unruly" vector of wildness as a positive force in human development (Merchant, 1994). Indeed, Francis Bacon is seen as the main figure responsible for solidifying the idea that the natural world must be dominated and controlled (Merchant, 1994). Thomas Berry (1999) noted that Bacon's ideas were received because of the existing climate of fear of nature and devastation following the ravaging global epidemic of Bubonic Plague during the 14th century, which impacted both Europe and Asia. And so, the major turn toward a penchant for dominating nature occurred in the aftermath of the Black Plague, which was so devastating that it engendered a sense of betrayal by nature.

Our yearning for control over nature is driven by our fears about natural disasters and disease but is also an aspiration to power (Leiss, 1994). For Nietzsche, this period's "progress", the discovery of absolute "truths", and its scientific and technical "innovation" were more about what he called the "Will to Power": the human drive to dominate nature and the environment (Leiss, 1994). Our "rational management" of natural resources is another manifestation of this will to power over nature. And throughout history, we see that power follows those who control these natural resources.

Plumwood (1993) traced Western culture's oppression of nature back to the construction of a "self" who is human, male and principally defined by his ability to reason. This ability to reason then became definitionally opposed to nature. In distancing the dominate human from nature, all that is associated with nature was also thrown into opposition, namely woman and her association with reproduction, emotions, and the body. This is where the similar discursive production of both women and nature have connected the two conceptually and symbolically. Eaton and Lorentzen (2003) maintain that Euro-Western cultures divided the world hierarchically and dualistically. These binary categories align women with the earth, the body, sexuality, femininity, nature, and materiality. Meanwhile, men are marked as masculine and conceptually connected with the mind, the supernatural, disembodied spirits, and heaven. And this ideological product has constituted both women and nature as having shared identification. These systems of power and the oppression crucial to their functioning, all differently predicated on the human, are understood as originating in the West (Boddice, 2011; Sanbonmatsu, 2011).

Dichotomous constructions such as reason/emotion, mind/body, culture/nature, heaven/earth, and man/woman give priority to the human, culture, reason, civilization, and the masculine. Ecofeminists have given a lot of consideration to these pairings, or better understood as hierarchical dualisms (Adams, 1994; Griffin, 1978; Haraway, 1989; Merchant, 1980; Plumwood, 1993; Walker, 1988; Warren, 1987). A number of ecofeminists have explored the ways in which such conceptual pairings point to a logic of domination that is deeply entrenched in Euro-Western history and contemporary worldviews. Scholars such as Jacques Derrida (2008), Marti Kheel (1995, 2008), Josephine Donovan (1990), and Cary Wolfe (2010), among many others, have argued that ontological dualisms such as human/animal, nature/culture, conscious/unconscious, and civilized/savage were never just deployed to devalue nonhuman animals, but also those marginalized persons precariously existing at the limits of humanity (Deckha, 2009). In such a scenario, appeals to reason are given more legitimacy over emotion. This value dualism then becomes a political tool, where women, people of color, nonheterosexuals, and others associated with emotion (or a diminished capacity for reason) are linked to animals. This value dualism is called a logic of domination and it can be used to operate other mechanisms of discrimination and oppression against human and nonhuman animals (Gruen, 2004).

A whole host of institutions, such as religion, science, philosophy, as well as cultural symbols and common discourse reinforce this worldview. Scientific knowledge is valued over indigenous ways of knowing. Commerce is valued over relationships. Abstract thinking is valued over the particular. These modes of dividing up aspects of our existence end up prioritizing one and denigrating the other. For ecofeminists, the two most crucial divisions: man, the masculine, is prioritized over women, the feminine, and human society and culture are seen as superior to the world of nature (Cudworth, 2015; Eaton & Lorentzen, 2003). The result is that male power over both women, Othered bodies, and nature appears "natural" and thus justified (Epstein, 2002; Gruen, 1993). We can examine a variety of social patterns that reflect this same template of domination, such as sexual norms, governance, neoliberal capitalism, education, and interpersonal violence.

The domination of the earth is a distinctly gendered phenomenon. Gender is linked to the management, control, and relation to the environment in a variety of ways. The increasing exploitation of the environment is rooted in the gendered character and orientation of Western culture. But it is also the case that decisions about environmental issues are made largely by men. Meanwhile, it is also the case that the people at the mercy of the outcome of those decisions are the most disenfranchised among us (Epstein, 2002;

O'Brien, 2017). While environmental crises might be described in generic terms, a gendered "lens" illuminates the ways in which such crises are shaped by the gender social order. For example, deep ecologists point to the growing human population and the coming calamitous impact it will have on the environment. What is often missed in such discussions is that population booms are very much connected to gender and oppression—a social reality of restricted access to reproductive technologies, beliefs about women's "place" in terms of child-bearing, and political climates charged with patriarchal values that impact the scope of our bodily integrity and reproductive freedoms (Flavin, 2009; Vance, 1997).

It is crucial to note that the origins of ecofeminist thought are from non-Western scholars. Indeed, ecofeminism is said to be "a new term for ancient wisdom", denoting that the roots of ecofeminist thought lie not in the West but rather in the spatiotemporal "elsewhere" (Mies & Shiva, 2014). In *Ecofeminism* (2014), Mies and Shiva discuss many of the same ecofeminist developments as Western scholars have but pay far more attention to ecofeminism as a survival strategy for persons in the Global South who deal with the devastating effects of Western colonialism and development every day. Mies and Shiva (2014) also discuss the appropriation of spiritual interconnectedness by Western ecofeminists, particularly in the USA. Indeed, while reaping the benefits of global inequalities, many posthumanist ecofeminists look to traditionally "Eastern" ways of being, thinking, and living. Throughout *Ecofeminism* (1993), Mies and Shiva (2014) explore a history of ecofeminism not centered in the Western world or institutions such as the U.S. academy.

The debate over ecofeminism "naturalizing" women

In the 1990s, ecofeminists explored the "othering" of women and animals by exposing their intersecting structures of power (Adams & Gruen, 2014). A misunderstanding over essentialism led to ecofeminism falling out of favor for a number of years before making a recent comeback (Warren, 2000). The controversy regarding essentialism takes issue with the conflation of women with nature and the early tendency to regard all women as a whole, without regard for class, race, age, nation, and so on (Thompson & MacGregor, 2017). After all, women oppress women in some contexts, so to regard all women as a singular oppressed entity is erroneous. More recent ecofeminist scholarship demonstrates more inclusion with regard to race, disability, sexuality, coloniality, and class, along with gender, when considering power relations and the environment (Gaard, 2017). Indeed, most environmental feminist scholars repudiate both essentialism and ethnocentrism and strive to understand women's relation to nature in a non-essentializing way (Thompson & MacGregor, 2017). Since this charge of essentialism in ecofeminist literature has had such an impactful history, and because the potential for misunderstanding the conflation of women with nature, I consider this issue in great detail next.

A key focus of ecofeminist theorizing has focused on how to conceptualize the link between women and nature (Adams & Donovan, 1995). Some ecofeminists maintain that patriarchy equates women and nature, so a feminist analysis is necessary if we are to understand environmental problems in their entirety (Griffin, 1978). This argument can be summed up as: the degradation of nature will also be accompanied by the degradation of women. It follows, then, that the practice of "feminizing" nature involves transferring attributes typically associated with women—such as weakness and nurturance—to the environment, rendering nature exploitable. Western culture is in many ways antinature, where culture is considered superior to nature, and women and Othered bodies are regarded as more "natural" or closer to "nature" and therefore inferior.

Most ecofeminist scholars take the position that the equating of women with nature is problematic and work to challenge and deconstruct this association (Evans, 2002; Gaard, 2004). Meanwhile, alternative spiritualities, or feminist spirituality, actually celebrate this connection through nature-based religions such as witchcraft, Native American spiritual traditions, and Paganism (Ruether, 1992). These religious traditions often celebrate images of female power and worship female deities (Ruether, 1983). These two positions are clearly at odds with each other. The first regards the patriarchal equation of women and nature as problematic; the spiritual position sees this equation as empowering. The extent to which women are celebrated as "closer to nature/earth" may, therefore, reinforce the idea that women should be confined to the world of nature, reproduction, etc., while simultaneously reproducing the notion that men are closer to non-natural things: culture, civilization, the human-constructed world, and reason. The tension here is that spiritually based ecofeminists have been criticized as reifying the fusion of women with nature with potentially harmful consequences (Warren, 2000).

Most mainstream feminists are wary of discussing any link between women and nature, for good reason. Feminist theory texts rarely include sections on ecofeminism. Women's studies conferences schedule very few ecofeminist sessions. Resisting an "ecological turn" in feminist studies in the academy is understandable. The conflation of women with nature is exactly the sort of reasoning that has been historically used to keep women in a position of subordination. In fact, many feminists work at dismantling myths of difference out of concern that too much difference will be used as a political tool against them (Bradley, 2013; Mohanty, 2003). The unintended consequence of avoiding discussions about gender and nature is that gender-ecology readings are left unexplored and opportunities for feminist interventions in environmentalism are missed.

Ecofeminists claim that women are disproportionately affected by environmental crises across the globe (Eaton & Lorentzen, 2003). This gender-specific outcome is primarily due to the gender division of labor that women are expected to provide subsistence for the family, and because gender norms may prevent women from owning property and resources that may assist them in times of crisis (Shiva, 1988, 2000). This claim that women and men experience a different relationship to their environment, and that women and men differ in their response to environmental issues, and even the claim that women are more responsive to nature, has resulted in ecofeminism being dismissed as essentialist (Zimmerman, 1994). These charges, however, are misguided. Indeed, most ecofeminists consider these connections between women and nature not as in essence but are rather socially situated and deeply embedded in our conceptual frameworks (Cudworth, 2005; Plumwood, 1993; Sturgeon, 2017). Where women have expertise in agriculture and ecological systems, for example, it is not due to their essential nature, but to their life experiences. Human experience is located in the social, not as some inevitable fixture in the body. Since these charges of essentializing women and nature, ecofeminist scholars have taken great care to avoid any misunderstanding that exploring the social connections between women and nature is in any way "naturalizing" this social relationship (Thompson & MacGregor, 2017).

As this section details, there are clearly some variations in the theoretical connections between women and nature, and ongoing debate about how to approach problematic connections, such as "women and nature". Yet overall, ecofeminism provides powerful analytical tools to investigate many areas of social life, environmental changes, and lived experiences. Moreover, ecofeminist theory makes use of a range of other feminist insights and draws on numerous feminist traditions. As Noel Sturgeon (1997) writes:

> Feminist critiques of forms of abstract rationality which reified divisions between culture and nature, mind and emotion, objectivity and subjectivity; psychoanalytic theories of the ways in which male fears of women's reproductive capacities structured male-dominated political and economic institutions; a feminist rethinking of Christian theology; critiques of the patriarchal nature of militarism; feminist anthropological research; feminist critiques of science; analyses of the sexual objectification of women; and feminist poststructuralist theories of constructed subjectivities are only a few of the vital feminist resources for ecofeminist theories.
>
> (p. 265)

In this sense, ecofeminism is not separate from other feminist modes of inquiry. Rather, these various strands of feminist thought are complementary, and ecofeminism has the potential to explain a great many social realities, including violence.

The scope of gender violence

Gender violence is a multi-textured phenomenon, shaped by the intersection of culture, geography, race, class, ethnicity, sexual orientation, age, physical ability, uneven relations of power, along with gender, in any given time and place (O'Toole, Schiffman, & Edwards, 2007). At a macro level, gender violence stems from culturally embedded assumptions of domination over marginal groups, and the general acceptance of violence as a means of maintaining such a hierarchy (Fawcett, 1996). Indeed, gender violence is normalized by a cultural climate that favors militarism, domination, war, and conflict (Merry, 2009; Mooney, 2000a; Price, 2005). These social conditions serve to feed attitudes of acceptance of violence as both inevitable and normal. At a more individual level, when seeking to identify what makes an individual more violable, the metaphor of "intersectionality" is typically employed to illustrate the cross sections of identity, oppression, privilege, and vulnerability (Crenshaw, 1991; Grzanka, 2014).

Gender violence takes on multiple configurations in varying social contexts. Gender differences, too, play out in a variety of patriarchal–heteronormative contexts. Whenever we see a situation where there is a pervasive privileging of masculinity over femininity—a societal dynamic which is necessary to the maintenance and reproduction of patriarchy—violence may be employed as a tool to uphold these asymmetrical gendered power arrangements (Lenton, 1995). Or, in the case of male on male violence, gender violence may be normalized through the compulsory expectation that boys and men display aggression as proof of their masculinity (Kimmel & Messner, 2004). In the case of violence against LGBTQ persons, gender transgressions are punished in an effort to preserve normalized gender performances (Everhart & Hunnicutt, 2013). Regardless of the particular identities of perpetrators and victims, violence is always experienced through the performance of gender. Violence involves the enactment of gendered beliefs (Throsby & Alexander, 2008). Violent performance reflects systemic beliefs that then reproduce gendered patterns of abuse. Whether we are considering the forced removal of genitalia to control a woman's pleasure or hyper-masculine displays of violent contests among prison inmates, violence performances are forged within frameworks where gender is a central organizing principle in the struggle for power (Kilmartin & Allison, 2007).

Gender violence is a widespread phenomenon which compromises the health and stability of communities and leaves survivors with lasting psychological and physical trauma. According to CARE, one in every three women has been beaten, raped, or abused in some other way—typically by an acquaintance, spouse, friend, or family member. Despite this pervasiveness, gender-based violence is one of the least-recognized human rights abuses in the world (CARE). Violence is a common tool of terror in conflict settings, with women often the main target (Tripp, Ferree, & Ewig, 2013).

Gender violence is not just about women, however. Violent victimization varies tremendously by key characteristics of the survivor. In fact, when looking just at homicide, aggravated assault, and robbery, women's rates of victimization are much lower than men's rates (United States Department of Justice, 1996). For rape, sexual assault, and domestic violence, however, women's rates of victimization are much higher than men's. The National Crime Victimization Survey (NCVS) estimates that about 90% of rapes are committed against females (Catalano, 2004). When it comes to intimate partner violence, the NCVS finds that women make up about 85% of all intimate partner violence victims (Rennison & Wechans, 2000). The National Violence Against Women Survey (NVAW) confirms this dramatically different rate, showing that women are far more likely than men to report serious intimate partner violence (Tjaden & Thoennes, 2000).

Results from the Conflict Tactic Scales (CTS) developed by Murray Straus and colleagues (Gelles, Loseke, & Cavanaugh, 2005), however, reveal a different picture, where they claim that there is actually gender symmetry in spousal violence. Using the CTS, participants are asked to report on behavior that ranges from very minor to very serious abuse, such as shoving each other to using a weapon. When asked in this fashion—on a continuum of abuse scales—results of CTS data show that men and women commit about the same amount of intimate partner violence (Gelles & Loseke, 1993). The "sexual symmetry" findings have sparked much debate between scholars (Gelles, Loseke, & Cavanaugh, 2005). Critics point out that rape is excluded from the survey, that men's violence against women is more severe than women's violence against men, and that CTS surveys remove context from violent events (Gelles, Loseke, & Cavanaugh, 2005). For example, it is possible that much of the violence women inflict against male partners is in self-defense, but the CTS are not designed to capture such scenarios (Dobash et al., 1992).

Generally speaking, with the possible exception of intimate partner violence, patterns of victimization for men and women look differently (Yllo, 1993). On the one hand, women are more likely to experience serious forms of domestic violence, rape, and sexual assault, and more serious forms of domestic violence (Stanko, 1985). Men, on the other hand, are far more likely to experience robbery, murder, and aggravated assault (Kimmel, 2006; Messner, 2016; Pinker, 2011). These differences are very much rooted in the gendered ways in which our lives unfold. Our heteronormative–patriarchal culture influences the degrees of vulnerability and protection for gendered beings.

Levels of violent victimization, as well as unique contexts of victimization, fall along gendered lines. Women are more likely to be killed by a heterosexual intimate partner than are men (Pridemore & Freilich, 2005). When women are murdered by an intimate partner, it often follows from a history of domestic violence (Dobash & Dobash, 1995; Moracco & Butts, 1998;

Tjaden & Thoennes, 2000; United States Department of Justice, 1996). Violence is, then, a product of gendered arrangements. When individuals are targeted in patterned ways, along the lines of race, ability, sexuality, gender, and so on, it suggests that they are being targeted precisely because of their presentation of self or demographics. It is also remarkable that when taking violence as a whole, men are more likely to commit violence and be the victims of violence (Pinar, 2001). That said, the vast majority of men are not violent and most of our lives are spent nonviolently most of the time, regardless of gender.

It follows that one of the sources of gender violence is the patriarchal structures of societies (Turner, 1998). Over the last 40 years, a substantial amount of theorizing has been done from a "gender center", of which radical feminists have contributed the greatest share of work. In the 1970s, radical feminists were the first scholars to advance the idea that patriarchy could explain male violence against women (Brownmiller, 1975; Caputi, 1989; Firestone, 1972; Griffin, 1971; Millet, 1970; Russell, 1975). Since that time, however, feminist theories in the areas of work, family, and sexuality have exploded (Chavetz, 1990; Chodorow, 1978; Oakley, 1974; Walby, 1990), while theory development on gender violence has stagnated. It may still be true that gender violence is "the most poorly theorized of all aspects of gender inequality" (Fox, 1993, p. 321).

A good deal of the early literature on gender violence failed at understanding men and masculinity as a set of complex social meanings where "manliness" is experienced differently by different demographic groups (Kimmel, 2006). Too often in gender violence narratives, men are included as an "unproblematic given", exempting men from careful consideration. In reality, gender is a central mechanism through which our life opportunities, privileges, and power are distributed (Price, 2012). Men contend with their own power and powerlessness depending on their social location. It is tempting to simplify the link between patriarchy and gender violence or to simplify male perpetrators as power-seeking, tyrannical abusers. Early work on violence against women sometimes conveys a dichotomous, "us versus them" scenario where women were controlled through abuse in order to reproduce the current patriarchal order and for men to exploit their position of advantage (Brownmiller, 1975; Caputi, 1989; Firestone, 1972; Griffin, 1971; Millet, 1970; Russell, 1975).

Now we understand that it is predominantly the least powerful men who victimize others and that their violence may manifest in reaction to social pressure to redeem their "wounded masculinity" and accrue more power (Kimmel, 2015; Messerschmidt, 1993; Messner, 2016). Dobash and Dobash (1979) elaborate: "Men who assault their wives are actually living up to cultural prescriptions that are cherished in Western society—aggressiveness, male dominance and female subordination—and they are using physical force as a means to enforce that dominance" (p. 24). Of course, even while all men receive similar cultural messages, most men do not aggress against women or other men. Moreover, patriarchal systems tend to operate with contradictory values of protecting women from violence, on the one hand, and valorizing male aggressiveness, on the other (Kilmartin & Allison, 2007).

We are all under the spell of hegemony that positions us to struggle for status within hierarchical systems. Gender violence between men is a manifestation of this gendered will to "superiority". Men are more likely to be the victims of homicide than are women. Indeed, men's behavior is forged within complex social structures (Fuchs, 2001) where violent behavior is a product to pathological social arrangements where human and nonhumans are sorted into categories of "worth" and where men receive cultural communication to strive for supremacy and authority.

Gaard (2017) points out the ways in which hegemonic masculinity is lethal. Mass shootings target women (as in the Montreal Massacre of women engineering students in 1989), Latinx and queers (as in the Pulse Nightclub shooting of 2016) and children (Sandy Hook Elementary School in 2012) and for the male shooters who usually perish by self-inflicted gunshot wounds or are killed by the police. These shooters are typically white males acting alone and the majority of these shootings (57% between January 2009 and 2014) involved the shooter killing a partner or other family member as part of the massacre (Gaard, 2017).

One feature of hegemonic masculinity is economic achievement among the dominant male. Marxist and Socialist feminists have considered the interplay between capitalistic and patriarchal systems and how these mutually reinforcing systems of domination affect gender dynamics in ways that result in economic domination taking gendered forms (Ehrenreich, 1976; Jagger, 1983). These deeply intertwined systems of patriarchy and capitalism create gender scripts which prescribe that males achieve breadwinner status. As men suffer economic hardship and alienation from hierarchical gains violence may be a response to these tensions. Gartner and McCarthy (1991) discovered that employed females who had an unemployed husband at home posed the highest risk for being killed. In this same study, the researchers found that the safest situation for married women was the reverse: to be employed with an employed husband. Economic insecurity arises from patriarchal scripts that cast men as dominant over women, other men, other species and over the earth. If men are under pressure to achieve, provide and protect, perhaps gender violence is more likely when dehumanizing labor conditions compromise these expectations.

Evidence on gender violence suggests that when men use violence, it is often to maintain their advantage in the most *disadvantaged* situations. Dorie Klien writes:

> Male physical power over women, or the illusion of power, is none the less a minimal compensation for the lack of power over the rest of one's

life. Some men resort to rape and other personal violence against the only target accessible, the only ones with even less autonomy. Thus, sexual warfare often becomes a stand-in for class and racial conflict by transforming these resentments into misogyny.

(Cited in Messerschmidt, 1993, p. 116)

The more disempowered men are from socially esteemed positions of dominance, violence becomes a way to achieve the only position of domination available (Connell, 1986, 1990b). When researchers consider men as perpetrators of gender violence (in heteronormative scenarios), many have concluded that the victimization of women has more to do with the status of males than of females (Baron & Straus, 1987; Brewer & Smith, 1995; Goetting, 1991). Those males who enjoy authority, advantages, and privileges are perhaps less likely to use violence against human others since their elevated position is sustained in more "legitimate" ways. Indeed, men are situated in their own scheme of domination relative to males and other groups not defined by gender. A focus on male power, then, must be situated within a larger hierarchical order.

It is more common to see mainstream theories employed to explain gender violence (Yllo, 1993). Theories that lack a gender-sensitive focus, however, end up obscuring the ways in which gendered power arrangements structure human action (Dobash et al., 1992). For example, the literature reveals that general systems theory, resource theory, exchange/social control theory, and subculture of violence theory may all be used to explain gender violence (Jasinski, 2001). Murray Stauss and Richard Gelles (Gelles, 1993) developed the "family violence" approach, which incorporates strands of each of various theoretical traditions to explain a whole host of violence we might see within families. The Strauss and Gelles (Gelles, 1993) theoretical models fail to centralize male dominance as a central, organizing feature, however. A similar problem can be found in individual-level theories that emphasize psychological dimensions of abuse. These perspectives may overlook the patriarchal climate in which violence plays out. In other words, individual level theories, in their focus on "sick" people, may miss seeing "sick" social arrangements (Flax, 1993). A microscopic focus on individual characteristics of violent actors presents a problem in that such an approach elides the larger social organization in which violence plays out.

Emphasizing the structural causes of gender-based violence, the Inter-Agency Standing Committee (IASC, 2005) takes the stance that "gender inequality and discrimination are the root causes of GBV". Meanwhile, the United Nations High Commissioner for Refugees (UNCHR) emphasizes the ideological causes: "[t]he root causes of sexual and gender-based violence lie in a society's attitudes towards and practices of gender discrimination, which place women in a subordinate position in relation to men" (UNHCR, 2003, p. 21) and "preventing sexual and gender based violence thus requires changes in gender relations within the community" (UNHCR, 2003, p. 35). While gender inequality, gender discrimination, and patriarchal practices are important elements to consider, there are other social realities that should be included in this inquiry.

Early work on gender violence lacked an intersectional perspective. As scholars and activists worked to raise awareness and craft policies and resources to help survivors, much of this scholarship played out in a decisively gendered and heteronormative context (Erbaugh, 2007; Goldberg & White, 2011; Loseke, 2001). LGBTQ survivors rarely fit into heterosexual and/or gendered categories associated with common gender violence narratives (Erbaugh, 2007; Everhart & Hunnicutt, 2013). Since most of the work done in the gender violence literature is about heterosexual couples consisting of cis-gendered individuals, violence among LGBTQ individuals has gone largely ignored. When looking just at intimate partner violence, same-sex couples likely experience violence at comparable rates to heterosexual couples (Erbaugh, 2007 Guadalupe-Diaz, 2013). This lag in recognizing LGBTQ survivors in academic research is also reflected in legislation. As of this writing, only three states (Hawaii, Maine, and Washington) specifically acknowledge gay and lesbian victims of intimate partner violence in their state's legislative code (American Bar Association, 2013). Other states explicitly define victims of intimate partner violence as heterosexual (Domestic Violence, 1989). LGBTQ victims of violence are omitted from research and policy even as the prevalence of violence among these populations is quite high.

The National Coalition Against Domestic Violence (NCADV) reports that 43.8% of lesbian women and 26% of gay men have experienced rape, physical violence, and/or stalking by an intimate partner at some point in their lifetime, as opposed to 35% of heterosexual women and 29% of heterosexual men. Similarly, and 61.1% of bisexual women and 37.3% of bisexual men have experienced rape, physical violence, and/or stalking by an intimate partner in their lifetime (NCADV). Race exacerbates risk where Black LGBTQ persons are more likely to experience intimate partner violence than those who do not identify as black. Non-normative gender identities and people of color are often left out of fields of concern when discussing gender violence, and our historical privileging of white female victims of violence has fed unfortunate cultural narratives about people of color (Pinar, 2001). Both people of color and economically marginal individuals are often associated with dangerousness and violence (Hollander, 2001).

Women of color occupy a place in our cultural imagination as highly sexualized, violent, primal, and aggressive (Newman, 1999; Ritchie, 2006; Wriggins, 1983). Meanwhile, black men, who endure more violence than any other demographic group in America, are typically portrayed as "dangerous others" (Holland, 2001). Rather than taking their victimization seriously, economically disadvantaged individuals are often labeled as deviant and are criminalized (Ignatieff, 1978; Piven & Cloward, 1993). Our unique assemblage of multiple identities shapes our social experience of risk, reward, safety and victimization (Collins, 2000; Crenshaw, 1991; Puar, 2007). Patricia Hill Collins (1991) articulates these interlocking systems of oppression where the mesh of an individual's oppressed and privileged identities impacts our life outcomes. There is a substantial amount of helpful research that examines gendered dimensions of violence (Bornstein et al., 2006; Renzetti, 1996; Walters, 2011), but far less scholarship considers multiple marginalized identities and violence jointly (Mendez, 1996; Ritchie, 2006).

Violent victimization falls along lines of ability as well. Our hierarchical values favor certain types of embodiments over others and particular bodies are presented as the ideal model against which all other bodies are compared. It is not surprising then, as Sunaura Taylor (2017) points out, that individuals with mental and physical disabilities have endured some of the worst violent victimizations. Disabled people are disproportionately represented in jails and prison, endure hate crimes, and are more likely than able-bodied persons to be victims of violence (Taylor, 2017). Historically, the atrocities directed against individuals who have mental and physical differences are terrifying. Violence against the disabled at the hands of institutions includes infanticide, eugenics, forced sterilization, and forced institutionalization (Taylor, 2017).

This review highlights a demand for a more elastic framework from which to understand gender violence; a reconstellation of the variables in this patriarchal field to include our material and ideological relationship with the environment. By linking environmental domination to gender violence, I am suggesting a fundamental overhaul of the gender violence framework, including its application by the global women's and environmental movements. What does gender violence have to do with domination of the more-than-human-life world? Patriarchal norms are reproduced by institutional practices, gendered performances, moral frameworks, and cultural expressions. Gender violence is inextricably tied to naturalizations, customs, traditions, and taboos that center on the natural world. And yet, it is the familiar patriarchal constructions of gender difference that elide the additional variables of environmental context. Both gender violence scholars and ecofeminism share common conceptual connections. Both expose the power-over-privilege systems of domination that inform basic interpersonal exchange.

What is nature?

While the term "nature" is a fixture in our cultural imagination, what does it really mean? The term is never neutral, but instead includes a variety of meanings with political implications. Depending on how it is constructed, it may be used to serve the interests of some groups, while disempowering others (Evans, 2002). Evans (2002) contends that the Western articulations of "nature", which were brought into the New World, conveniently worked to preserve the interests of hegemony. Indeed, nature is a social construction which is often crafted in such a way as to serve particular ideological interests. The word "nature" is derived from the Latin root meaning "born". The use of the term "natural" in this context refers to a behavior that is automatic for a species, such as swimming for whales, roosting for hens, and walking for humans. When used this way, the term is essentialist in the sense that any "natural" attribute is considered organic, inborn, and therefore largely beyond our ability to change. To attribute any behavior to our "nature", take rape, for example, is to normalize the behavior as inevitable and innate. When any human activity, action or behavior is labeled "natural", such behavior may then be used to justify domination (i.e. the belief in the "natural" inferiority of women and people of color). Mei Evans (2002) elaborates:

> When it is said that women are "by nature" maternal, that people of color are "naturally" more in tune with nature, or that it is "unnatural" for people of the same gender to be sexually attracted to one another, what role is being assigned to nature? What is the work of culture, of human-constructed relations, that nature is being asked to perform in these equations? What is at stake for these groups of human beings, and what is at stake for nature itself?
>
> (p. 184)

Racist and sexist ideologies often include a reference to "natural affinities", "categories authorized by nature, destinies inscribed in biology, and 'scientific proofs' of the limited capacities of the 'other' that have rumbled through the centuries to justify slavery, the oppression of women, and ethnically and racially based holocausts and genocides" (Seager, 2003). Consider racist beliefs that assume that the observed inequality between whites and people of color reflects "given" capacities (Salleh, 1997). Sexism that draws on beliefs about women's reproductive process draws conclusions that women's intellectual, social, and emotional functioning should be limited to nurturing and caregiving (Salleh, 1997). In a hierarchical arrangement where nature is subjugated, the social division of labor between men and women may be regarded as normal, even inevitable, because it springs from women's "natural" inferiority. Salleh (1997) goes on to point out that "arguments for the slavery of dark-skinned people echo the same logic" (p. 37).

It is therefore impossible to separate our own biases from our definition of nature. In other words, nature is never separate from human thought. Yet, to shore up support for a variety of arguments that lay claim to certain human behaviors as "natural", we often look to behaviors in the nonhuman animals' world. But this practice of pointing to nonhuman animal behavior as evidence for the "natural" origins of our own is particularly prevalent in heteronormative discourse. Consider the claim that queer sexualities are "unnatural". This argument is flawed since behavior in one species cannot be reliably applied to the norms and behaviors of other species (Gaard, 2004). Any oppressed identity group is seen as "closer to nature", in the dualisms and ideology of Western culture. However, it is often said that queer sexualities are devalued for being "against nature" (Gaard, 2004).

In 2010, a New York Times Magazine story titled "They Gay? There is a Science to Same-Sex Animal Behavior" explored whether or not animals are gay. The author of this piece pointed out how much we tend to base human activity on animal activity. The author, Jon Mooallem (2010), dismisses this by pointing out that humans do all sorts of things that animals don't (e.g. sit at tables and talk, and play scrabble). It turns out that we typically look to the animal world for reassurance of our own human behavior when it comes to issues of sex.

In his study of the bonobo (pygmy chimpanzee), a species that, together with the chimpanzee, is the nearest relative to *Homo sapiens*, Frans de Waal (1995) found that sexual behavior served a variety of reproductive and nonreproductive functions. In effect, research on nonhuman primate sexual behavior indicates that for nonhuman primates much of their sexuality is nonreproductive. I use the example of "naturalizing" normative sexuality by pointing to the more-than-human-life world to explore the range of this rhetorical technique. We also see these same rhetorical devices used to naturalize violence.

Naturalizing violence

Ecofeminists have critiqued the essentialist equating of women and nature, as well as the tendency to dismiss the behavior of men and women as "natural" (Sturgeon, 2017). Violence is often essentialized and "naturalized" in ways that parallel the ideological processes that "naturalize" women. Such essentialist ideological justifications lead to a depoliticization and inevitability of violence, which in turn informs the architecture of domination, hierarchy, and power arrangements. A widespread belief about violence is that it is a naturally occurring phenomenon in both human and nonhuman animals. This argument might be bolstered by pointing to the no shortage of examples of bodily harm and killing throughout the species-verse. Moreover, certain ideas and beliefs about what is "natural" might be deployed to justify gender violence. Indeed, even hierarchies seen in nature may be used to legitimate the violent maintenance of human hierarchies (predator and prey, for example).

Historically, essentializing rhetorical arguments that appeal to "nature" have been used as tools to justify all sorts of exploitations—from violence against indigenous peoples to the commodification and abuse of animals

(Harper, 2011; Pinar, 2001). Pointing to violence within the nonhuman animal world serves as thin evidence to legitimize other brutalities—as if there is an intrinsic will to kill and that the promotion of nonviolence is "going against nature". Of course, the way we construe "nature", "unnatural", and "natural" ways of being is deeply intertwined with religion, politics, specific histories, and cultural mythology (Merchant, 1994). To the extent that violence is considered "natural", it is actually derived from our ideological relationship with the other-than-human-life world. The "violence is natural argument" is gendered, and it stems from a particular ecological and patriarchal worldview. In this instance, the term "natural" is used to indicate accepted standards of behavior and is used to justify domination. Naturalization of violence among men, in particular, is common. Violence, or aggression, among men, is routinely regarded as innate and inevitable. Evidence to the contrary is ignored because of natural assumptions and the persistence of Darwinian ideas that ignore partnership and cooperation.

Beliefs about "natural" tendencies toward violence are not always rooted in biology, but also religion, where particular attributes are believed to be instilled by a creator. Gender bias is deeply entrenched in ancient religious texts—documents that also lay claim to "natural order" through creation myths and codes of conduct ordained by "God". Islamic law, or Sharia law, is a penal code that dictates punishments aimed at preserving patriarchy by criminalizing adultery and homosexuality. The Hebrew bible condones gender violence in many instances. In 2019, the small Islamic country of Brunei implemented a law that dictates death by stoning for adultery, abortion, and sex between men. Under this harsh law, lesbian sex is to be punished with 40 lashes (NYT, 2019). In these cases, violence is justified as part of a "natural order" established by a creator, thus perpetuating the naturalizing and normalization of gender violence. Religion is not the only context in which we see an appeal to nature as justification for violence. Historically, conservative entities have shored up support for their various power structures by essentializing and depoliticizing human behavior. Our understanding of nature is deeply entrenched in patriarchal and anthropocentric paradigms—where our fundamental orientation is one of human domination over nonhumans and othered humans. This human domination extends to the biosphere as well.

We often attribute certain human activities to our "natures" for reassurance and justification for a wide range of behaviors. What is often overlooked, however, is that the "violence is natural" argument is in fact derived from our ideological relationship with the other-than-human-life world. The belief that violence is natural also stems from a gendered, particular ecological and patriarchal worldview. For example, Eriksson Baaz and Stern (2013), in their study of sexual violence during war, point out that there has been a shift away from articulating sexual violence as the result of biological drives to understanding rape as a "weapon of war". In other words, there has been movement from what Eriksson Baaz and Stern (2013) call the "sexed" story to the "gendered" story. This gendered reading of violence stands in opposition to reading violence as "natural". But this is a recent shift. In other times and places, rape has been articulated as "an innate drive".

Violence is a part of our reality, of course. Even most nonviolent activists and theorists acknowledge that there is a time for violence. While there are societies at peace, as well as anthropological evidence of mostly peaceful societies in our distant past, a completely nonviolent world doesn't seem entirely possible, although most of us live nonviolent lives most of the time. And while there certainly is violence in nature, the narratives surrounding violence among nonhumans often involve crude observations deployed alongside simplistic beliefs about how species consume other species. This consumption narrative, which takes violence as its key plot line, tends to overwhelm any other interpretation about relationships in nature. Bekoff and Pierce (2017) lament that the

> consumption paradigm … has monopolized discussions of the evolution of social behavior. The predominance of this paradigm in ethology and evolutionary biology is both misleading and wrong, and momentum is building toward a paradigm shift in which "nature red in tooth and claw" sits in balance with wild justice.
>
> (p. 45)

While violence is certainly a part of human life, our cultural preoccupation with violence suffers from a fallacy of overgeneralization. In rendering violence inevitable, we tend to overlook the interplay of culture and nature, or the fact that nature is neither fixed nor unchanging. Steven Pinker, in *Better Angels of Our Nature* (2011), points out that there has never been another time in human history when there have been so few violent deaths. And yet "the violence as survival narrative"—who wouldn't use violence in the face of threat?—seems to have gained cultural currency, particularly in xenophobic and anti-immigration rhetoric. Such stories about using violence to survive as a part of our "nature" are also used to justify the slaughter of animals, even in cases when doing so is unnecessary. And while there is brutality in nature, there is also care, play and cooperation in a nuanced biotic community. We all recognize the potential for violence within us. Social psychologists have shown that most anyone is capable of anything (including violence of the most brutal sort) under compelling enough circumstances (Zimbardo, 2009). And yet, given the PTSD that war veterans face, one could argue that violence is very unnatural—devastating in its psychological effect.

Masculinist approaches to the other-than-human-life world can infect our understanding of our own behavior. If we interpret nature as a fierce and barbarous place, then we might look to nature to lend justification to all sorts of violence—against each other, against the earth, against animals as just "how we are". In addition, a favored cultural narrative involves the location of humans at "the top of the food chain" as a consequence of our "natural" positioning. Such "nature" ideologies that include fantasies of dominance are similar to ideologies of dominance in the practice of gender violence. This "logic of domination" over the environment makes male power over women and marginal groups appear "natural". Further, violence by males is often "naturalized" by appealing to beliefs about the other-than-human-life world. This fallacy links masculinity and its corollary of dominance, aggression, hierarchy on the one hand, and mastery over earth on the other hand. Clearly, there are several parallels between the way humans relate to the environment and the nature ideologies that we are embrace. But one consistent theme in popular nature narratives is the centrality and dominance of human animals and the need to maintain that advantage by keeping nature in check.

"Taming" nature

Nonhuman animals represent complex symbols for human animals—everything from "man's best friend" to vermin that need to be extinguished. While humans project a range of attributes and values on animals, one consistent cultural proclivity is to subdue the nonhuman. For Darwin and his contemporaries, "taming" nature and domestication was the path to "civilization", a gendered process that involved managing the "animal" character within (Knoepflmacher & Tennyson, 1977). Evidence of this practice of disciplining wild entities abound. While we might regard some animals as emblems of the "wild and free", others are subject to our hatred, captivity, manipulation, experimentation, and domesecration (Torres, 2007). Through elite horseback riding, dressage, the elegant triumph of "civilized" humans manage and control a formerly "unruly", but now "broken" nonhuman animal in choreographed pageantry. Piñatas are usually in the shape of animals. We beat them with sticks until goodies pour forth—a ritual reserved for special occasions and celebrations. Consider the gratuitously violent display of animals in rock concerts, such as when Ozzy Osborne bit the head off of a bat during a live concert, communicating that animals should fear humans and that their violent ending serves as spectacle and entertainment. The iconic statue of a charging bull located at the Wall Street stock exchange—the headquarters of global capitalism—serves as a reminder that great wealth is procured from conquering even the most formidable creatures in the natural world. We may admire apex predators for their physical tenacity, perhaps longing to emulate their hunting skills or elevated placement in a pack or nature hierarchy—and exterminate them for competitive compulsions (Emel, 1995).

Val Plumwood (1996) in her essay on Prey to a Crocodile tells a survival story that illuminates our conflicted positions with regard to animal others. Plumwood feels outraged that she would be considered food because, after all, she is human. While the crocodile attempts to kill her, Plumwood (1996) realizes that since humans regard ourselves as above "the animal" we, therefore, do not position ourselves as prey. And yet, while we don't consider ourselves prey, we also don't think of ourselves as predatory animals, despite practicing the routine, normalized, institutionalized, globalized mass killing of nonhuman animals. The most common relationship we have with domesticated nonhuman animals is that we eat them, and this requires the routine killing of enormous numbers.

Even in places where there is an apparent respect for nature, such as the wilderness, we can tease out a hidden colonization of nature—a taming of what is wild. Wilderness is idealized as "perfect nature" when it is devoid of humans; untouched by humans. This notion leads to a hyper-separation of culture and nature. Just as certain forms of culture are considered low and high, certain forms of nature are considered low and high. Humans live as if they are independent of and impervious to the natural world and human ingenuity can overcome all of nature's challenges (Vance, 1997). Wilderness supports whole industries (commercial backpacking tours, etc.), but wilderness is also governed by a whole host of bureaucrats who decide how the land will be managed and controlled (Vance, 1997).

Gender violence through metaphor and language: a gender-ecology reading

It is possible to observe the ways in which gender, violence, and nature are connected conceptually and symbolically. Social phenomena materialize "through the symbolically constituted and linguistically mediated encounters and interactions through which meanings and representations are communicated from one mind to another in the course of human association" (Scott, 2010, pp. 16–17). There is, therefore, an intimate congruity between language and social structure, where our linguistic configurations often reveal a view of nature as a threatening force. Symbolism then becomes a crucial tool in this work. We can learn a lot from the ways that humans relate to their environment through symbolic analysis. Our beliefs about how humans are situated in a hierarchical order relative to the-other-than-human-life world are evident in animal metaphors and idioms (e.g. riding the bull, bullfighting, dog and his "master"). These metaphors often involve violence to maintain dominance. The phrase "dog-eat-dog-world" conveys not only a distrust of nature but also that we regard animals and nature as cruel. Conversely, expressions of human-to-human violence are almost always expressed and understood through nature metaphors. Consider how often we seek to understand violence by associating the actors with nonhuman nature, such as an animal, beast, alien, dog, or pig. It is tempting to dismiss language and imagery as harmless comparisons, but of course, discourse, in its myriad of forms, is political. Words matter because language transmits and replicated attitudes, beliefs, behaviors, and practices.

The maltreatment of marginalized human populations is often legitimated by making disparaging associations between abject human Others and animality (Oliver, 2012). The phrase "I'll tan your hide" fuses violence against humans with historic processing of leather—where a cow is violently subjugated, its skin used as commodities for human use. Consider also the phrase "tail between his legs", a mechanism of shaming men for their cowardice by associating them with dogs. These animal metaphors are rooted deeply in speciesism, but also contain stereotypes based on a profound lack of understanding about the complex lives of nonhuman animals.

In language surrounding violence, we can readily observe the simultaneous domination of women and the domination of nature and how these images are often conflated with violence. We frequently observe the deployment of pejorative "gender-nature imagery" in violent acts where the dehumanization of women is accomplished by association with a particular type of animal. Women are frequently animalized through vituperative epithets (birds, queen bee, old crow, chicks, bunnies, whale, cow, bitch, sow), and animals are feminized when they serve as the insult to which "woman" is fused. The association with the nonhuman animal almost always involves negative traits. It is not uncommon for both animals and women to be rendered as frail, deficient, weak, and dumb. Even though animals are complex beings with varied capacities for intelligence, they have lodged in our cultural imagination as deficient. Rooted in both sexism and speciesism, the symbolic association between women and animals bolsters their exploitation. Such examples expose how speciesism uses sexist logics to function. The two associations of women and animal with deficiency and dehumanization should be continually and consistently challenged.

To commit violence, we suppress those traits in us typically associated with femininity (empathy) and degrade our victim by associating them with ruined women (son of a bitch). A failure to fight earns the label of coward—and cowards are jointly disparaged as feminine (pussy, wimp), and animals (dogs, beasts, devils, pigs). Insults toward humans summon our contempt for animals (calling a person an ass, snake, dog, hare-brain, or bird-brain). Nonhuman animal comparisons that are used as insults (e.g. monkey, gorilla, filthy as a pig, bullheaded) also serve to sustain exploitation and abuse of animals.

The human ability to devalue and distance human and animal Others suppresses empathy and invites contempt and detachment—conditions which make violence possible (Gay, 1993). Indeed, great atrocities throughout history are linked to devaluing the other through "animalization". Consider episodes of genocide throughout in history where humans were animalized (Rwandan genocide Tutsis were called cockroaches; In Nazi Germany, Jewish people were regarded as rats). Wolfe (2013) reminds us that in such situations humans are all potential animals. While violence against human animals is generally prohibited, nonhuman animals are subjected to "originary violence" and a "violence of sacrifice". The human/animal distinction is reproduced again and again when animals are regarded as sacrificeable while humans are usually not (Wolfe, 2013). Sanaura Taylor (2017) details degrading comparisons between animals and intellectually and physically disabled people. These comparisons range from insulting to legitimating racism to a complete loss of autonomy and freedom.

One early feminist works examined the parallel between animal and human slavery: Marjorie Spiegel's *The Dreaded Comparison* (1996). And in 1990, Carol Adams's book *The Sexual Politics of Meat* explored the comparisons between sexualized women and animals. A. Breeze Harper (2011) reminds us that pro-slavery whites deeply believed that Africans were "just like animals", could not feel pain, and were believed to have no spirit, soul, or even feelings. She writes:

> Growing up in the little American town of Lebanon, it was difficult for me to argue with peers and teachers about the importance of addressing racism and whiteness. Simultaneously, I felt isolated and frustrated by speciesism that was also accepted as the norm, and which surrounded me daily. Neither peers nor teachers understood why I refused to participate in dissection, and why I did not "appreciate" deer hunting (a huge "sport" in my town). After I told my fifth-grade teacher that I didn't want to drop a live worm into alcohol to kill it, and then dissect it, he told me repeatedly that worms do not have central nervous systems; hence, they "do not feel pain." Only through repeated stories, in my household, which exposed how our people were treated, did I become fully aware that pro-slavery whites deeply believed that Africans could not feel pain; that we were believed to be "just like animals" who had no feelings, spirits, souls; we were just machines available to serve the purposes of white America. Perhaps my fifth-grade teacher did not know this.

(p. 74)

The "animalization" of humans has led to colonization, a painful history of racialized violence, interpersonal violence, and violence against other species. Do we run the risk of reproducing speciesist discourses and dehumanization of Others by engaging in an analysis that examines the degrading comparison between animals and humans? PETA (People for the Ethical Treatment of Animals) has been heavily criticized for their ads which juxtaposed human suffering alongside animal suffering, often in racialized contexts, such as composing images that include Native American's on the Trail of Tears marching alongside animals being led to slaughter, portrayals of crowded Jewish concentration camps vis-a-vis farm animals crammed inside warehouses on a factory farm, or images of a lynched body juxtaposed with the body of a charred animal carcass (Harper, 2011). These sorts of comparisons often elicit outrage for implying that certain groups of people are equated with animals. And what about the animals? In these scenarios is speciesism left unchallenged? In *The Dreaded Comparison* (1996), Marjorie Spiegel writes:

> Comparing the suffering of animals to that of Blacks (or any other oppressed group) is only offensive to the speciesist: one who has embraced the false notions of what animals are like. Those who are offended by comparison to a fellow sufferer have unquestioningly accepted the biased world view presented by the masters. To deny our similarities is to deny and undermine our own power.

Sunaura Taylor (2017) asks the question "How do those of us who have been negatively compared to nonhuman animals assert our value as human beings without either implying human superiority or denying our very own animality?" (p. 110). Taylor (2017) goes on to discuss claiming animality, reminding us of the fact that we are all animals, but also that our oppressions as human animals and nonhuman animals play out differently. Claiming animality, even celebrating it, as Taylor points out is also connected to privilege and Whiteness. Some of us may do so more easily than others. But one thing is clear, as long as speciesism is left unchallenged it will remain a discourse available to others to leverage violence against other species (Wolfe, 2010).

So how do we challenge speciesism, reassert the animal, claim our animality without also reinforcing the dehumanization of Others? Taylor (2017) challenges us to approach these issues in new ways:

> It is undeniable that animals have experienced terrible violence at the hands of humans—violence that very often shares a genealogy with the violence humans inflict on one another. What if we saw the terrible acts they have suffered as an example of why they deserve not only our empathy and respect, but also our acknowledgement that they are our kin? What if instead of demeaning us, claiming animality could be a way of challenging the violence of animalization and of speciesism—of recognizing that animal liberation is entangled with our own?
>
> (p. 110)

Rape of the earth

When the phrase "rape of the earth" is evoked, the earth is transformed into a consumed and violated female body. When nature is described, it is feminized (mother earth) (Warren and Wells-Howe, 1994). As such, this phrase declares a symbolic connection between nature and women but it also references women's experience of sexual violence. It also suggests that nature is a place to fear—a site of rape. To the extent that "nature" is a place for people to experience recreation and enjoyment, those who are socially identified at "Other" may experience fear and even risk in this space so firmly equated with the hegemonic male. Consider the fusion of gendered and racialized fear of violence which is also bound up with nature. In "Black Women and Wilderness", Evelyn White explores her experience of being excluded from nature out of fear, particularly its recreational opportunities, from her own standpoint of being an African American female. While participating in a summer writer workshop in a wooded setting, White (1995) shares her fear and unease with being in nature:

> I stayed holed up in my riverfront cabin with all doors locked and window shades drawn. While the river's roar gave me a certain comfort and my heart warmed when I gazed at the sun-dappled trees out of a classroom window, I didn't want to get closer. I was certain that if I ventured outside to admire a meadow or to feel the cool ripples in a stream, I'd be taunted, attacked, raped, maybe even murdered because of the color of my skin.
>
> (p. 378)

White goes on to link her fears to memories of racial violence, particularly her awareness of her ancestors being violently hunted down in rural places "exposed, vulnerable, and unprotected—a target of cruelty and hate" (p. 378). Indeed, acts of violence against "Othered" bodies in rural or natural settings are familiar. The NAACP reports that of the 4,743 people in the USA who were lynched between 1882 and 1968, 3,446 were African Americans and more than 73% of lynching in the post-Civil War period occurred in the Southern states. The predominately rural settings conjure up images of trees as in Billie Holliday's famous song, Strange Fruit—a metaphor binding a tree's fruit with lynching victims. Recalling this terrifying historical fusion of violence against blacks with nature, White (1995) writes: "In (Emmett Till's murder) I saw a reflection of myself and the blood-chilling violence that would greet me if I even dared to venture into the wilderness" (p. 380).

While the "rape of the earth" is typically deployed to demonstrate a violent and dramatic taking of the earth's resources for human consumption, it simultaneously conjures our rape culture. The 1998 dragging death of James Byrd Jr. in the rural area of Jasper, Texas, and Matthew Shepard, a gay college student, who was beaten to death and left to die in remote, rural areas of Wyoming serve as reminders that nature can be sites of violence toward "Othered" bodies. "Other" here is anyone outside of the perceived norm of male, "civilized", cis-gendered, heterosexual. Women, differently abled, racial and ethnic minorities, gender nonconforming, homosexual, and indigenous people are all "Othered". The history of terror against non-white, nonheterosexual individuals seems to allot the "wilderness" as a cultural space reserved for "men in nature". Wilderness appears to be a hetero-masculinist social construction that is both racist and sexist where we see the policing of the boundaries of wilderness with the threat of violence against "Others". To challenge the hegemonic "ownership" of nature enforced by fear, White (1995) suggests that blacks and especially black women to be able to belong to the wilderness without fear and that situating themselves in natural settings is one part of resisting race- and gender-based oppression.

In the 2006 documentary, "Jesus Camp"—a documentary about a summer Christian camp for children which advanced teachings centered on the fusion of nationalism and religion—one of the camp residents was recorded passionately encouraging the audience to:

> rape this world … rape this earth, take everything you want from her … because, you know what? It doesn't matter. We are not here for very long. Christ is coming to take us away from Earth … so cut down our trees, use all of our oil … take advantage of everything that the Earth has to offer.

This quote, while an admittedly extreme example, exposes a long and pervasive believe about a "natural order" rooted in a history of religion and colonialization where human domination prevails, a violent occupation where the earth is a feminized entity to be "raped" for the consumption and benefit of humans. An entitlement to this violent procuring of the earth's resources seems to be a consequence of our "natures". The freedom to "rape" without consequence (it doesn't matter) and the promise of being airlifted from the planet (Christ is coming) betrays a conservative stance found in neoliberal politics of pursing profit with little regard for the damage that results from our actions (Chomsky, 1999). This rampage over the earth evident in this quote is fused with imagery of violence against women (take everything you want from her) (take advantage of all she has to offer) revealing that this imagined hierarchical utopia is rooted in patriarchy thinking.

The phrase "rape of the earth" conjures up images of the joint violent conquest of land (earth) and its native inhabitants. Conquest is carried out mainly by groups of men, bringing with them new concepts of masculinity. Rape and other forms of gender violence are a common feature of conquest. Smith and LaDuke (2015) explore the connection between the exploitation and colonization of native people bodies—especially women's bodies—and native lands, emphasizing patriarchal ideologies that hold a joint disregard for nature, indigenous people, and women's bodies. Smith and LaDuke (2015) point out how colonizers justified the appropriation of native lands by claiming that indigenous people did not subdue nature. The key ideological tool in this process is patriarchy:

> Thus, in order to colonize a people whose society was not hierarchical, colonizers must first naturalize hierarchy through instituting patriarchy. Patriarchal gender violence is the process by which colonizers inscribe hierarchy and domination on the bodies of the colonized.
>
> (p. 23)

As native people were marked as inherently "violable" by colonizers, Smith and LaDuke (2015) consider the various ways that native bodies were violated as land their land was claimed and misused. Toxic waste, the effect of which can manifest as illness, is far more likely in communities of color (Checker, 2005; LaDuke, 1993). American Indian lands, in particular, are subject to environmental racism where uranium production, waste dumps, military, and nuclear testing leading to leaked radiation into the atmosphere often take place on native lands (Smith & LaDuke, 2015). The blatant disregard for "the health and culture of indigenous peoples (through environmental destruction) is intimately connected with the destruction of the forests and land that constitute their homes" (Warren, 1990, p. 6).

Environmental racism inflicted on indigenous populations has violent consequences, namely the "ecocide of Native America" (Grinde & Johansen, 1995) and state-sanctioned environmental violence (Kuletz, 1998). Since gender-based violence is bound up with other human rights violations, indigenous women are often rendered more violable. Indigenous women in Australia are 45 times more likely to experience family violence than other Australians, a condition which is linked to a legacy of racist subjugation and the forced removal of children from their families in the 20th century (Susskind, 2018). These same communities continue to be subject to institutionalized racism, lack of health education programs, employment, and housing (Susskind, 2018). The consequences are far-reaching:

> First, concentrated and localized environmental hazards do not simply harm individuals, they erode family ties and community relationships. An onslaught of miscarriages or birth defects in a neighborhood, for instance, will create community-wide stress that will debilitate the neighborhood in emotional, sociological, and economic ways. To ignore this communal harm is to severely underestimate the true risk involved. Second, because concentrated and localized environmental hazards tend to be unevenly distributed on the basis of race and income level, any resulting mass injury to a threatened population takes on profound moral character. For this reason, Native Americans often characterize the military's poisoning of Indian land as "genocide".
>
> (Verchick, 2004, p. 74)

Individual cases of gender violence, particularly in indigenous communities, are rooted in a larger matrix of problems. A 2006 report from International Indigenous Women's Forum (FIMI) notes:

> Violations of collective rights have also subjected indigenous men to armed conflicts, environmental destruction, displacement, migration, urbanization, racism, unemployment, and poverty; and have exposed indigenous men to drugs, alcohol, models of masculinity predicated on domination, and religious doctrines that sanction male violence.

Conquest is, of course, fundamentally about the acquisition of wealth. The phrase "making a killing" exposes how violence is inherent in the pursuit of profit. Torres (2007) explores how violence is intrinsic to the social relations that uphold industrial civilization where "…for profit, we will do just about anything. We will destroy the natural world to the point of no return. We will force people to labor for us. We will kill and consume animals in the billions" (p. 68). The exploitation of socially marginal, powerless groups of humans and nonhumans are caught in the violent dynamics of conquest. Industrial civilization and neoliberal capitalism value the property of those at the top of the hierarchy over the human lives below and the nonhuman lifeworld from whose labor and resources are extracted for production (Torres, 2007). Cultural phenomena and economic factors collude in these scenarios. In other words, through a complex dialectic, conditions of structural inequality privilege the few and harm the many.

Conclusion

Environmental feminists explore the parallels between the social organization of gender and the ways in which societies are organized with respect to "nature"—pointing out, in particular, the joint devaluation of women, Othered marginal groups, and the earth. This text makes a case for the inclusion of a gendered reading of ecology in the analysis of gender violence. Gender violence and our relationship to "nature" are interrelated social phenomena and are informed by similar ideologies and structural arrangements. This text extends the environmental feminisms literature to illuminate the problem of gender violence. This work brings into dialog a range of disciplinary voices: gender violence scholars, intersectional movements within feminist studies, green criminology, ecofeminism that draws on the work of indigenous feminists (Kwaymullina, 2018), environmental justice scholars in the global south (Shiva, 1988) and feminist of color (Davis, 1983; Mohanty, 2003).

We find feminist environmental scholarship in a whole host of disciplines—an array of disciplinary approaches make use of this mode of analysis. Here I tease out the intersection of gendered degradation of the environment and gender violence. In addressing environmental injustice and gender violence simultaneously, I point to a set of common denominators shared between these two issues. Gender is missing from most environmental fields of inquiry and also from mainstream violence studies. Leaving gender out of our attempts at understanding violence, means that any effort toward correction, intervention, conceptualization, or examination will leave us with inadequate solutions and partial understandings. Because ecofeminism promotes an awareness of interdependence and a practice of nonexploitation, ecofeminist theory has the potential to generate insights and recommendations for many social problems (Warren, 1990), including violence. This book rests on the contention that we need to incorporate ecopolitics into the way we think about gender violence. It is crucial that we continuously examine the shared systems and ideologies that feed both the destruction of nature, particularly violence against nonhuman animals, and gender violence. Nature is a feminist issue (Warren, 1996, p. 1) and understanding it as such reveals the subordination of Othered groups.

Assumptions about human violence reflect our beliefs in the nonhuman life world. These assumptions are deeply gendered. This "naturalization of violence" is a belief that violence among humans is an essential part of the human condition. We point to the violence that occurs in the nonhuman life world as evidence that human behavior follows a parallel process. Nature ideologies are deployed to justify killing animals and exploiting the earth. One example is the widely held belief that it is part of a natural order, or food chain, to kill animals for human consumption. Ideas of "nature" and "naturalness" are socially constructed and are bound up with other ideologies of evolutionary fitness, normalcy, sexism, sexuality, independence, and so on.

Gender serves as an organizing principle through which to view violence. Gender violence and ecological degradation cannot be understood without a feminist analysis because these twin dominations are outgrowths of similar patriarchal structures. This framework puts an emphasis on violence as a gendered phenomenon and our relationship with the earth as a gendered process; yet, these are regarded as two parts of a shared reality. We must pay attention not only to inequalities within human hierarchies but also the cultural practices that involve exploiting the earth.

Human–animals' interaction with the environment has resulted in uncommonly destructive consequences. There are parallel trajectories of common oppressions of violence between humans and destruction of the natural world. These common oppressions occupy the same space in a culture of brutality that is distinctly gendered. These oppressions are culturally analogous and interdependent. I am referencing patriarchal culture where even protective responses reflect masculinist orientations that exclude empathy (Kheel, 2008; Kibria, 1990). Certain cultural mythologies serve to naturalize representations of violence against the earth in order to reproduce anthropocentric values. Words reproduce the ways that we treat each other. Representations, images, metaphors, and cultural narratives have social and political consequences for both humans and nonhumans. Such rhetoric sets us up for the risk of lived experiences being shaped or even replaced by stereotypical cultural narratives.

Understanding, solving, and analyzing the problem of gender violence should involve unearthing the power dynamics behind the scenes. If we include attention to plants, animals, land, fungi, and water, could we gather insights into patterns about social relationships, in particular, the twin dominations of marginal social groups and nature? The approach outlined in this text could be used in examining specific cases of gender-based violence. But this text will hopefully open up additional lines of inquiry related to the intersection of gender violence and domination of the earth. For example, as climate change progresses, how do the effects of environmental degradation result in violence? And how does that violence fall along gendered lines? Answers to these questions involve applying feminist thinking to the relationship between gender violence and environmental problems. The work to connect gender politics to the politics of environmental justice in more recent work has opened up new avenues of inquiry and analytical possibilities (Di Chiro, 2006; Mies & Shiva, 2014). This text takes the central analytic of ecofeminism as a powerful tool to understand gender violence, centering nature, but also valuing voices from the margins. Ultimately, this text concludes that the work on gender violence is not separate from the work on a green future.

Explaining violence is a very thorny and challenging process, yet insights of ecofeminism can be used to understand this social problem. It is my hope that such an "ecological turn" in the study of gender violence will widen our theoretical field of vision—expanding into new literature, drawing on cross-disciplinary ideas—to deepen our understanding of complex and gendered terrain of violence. Indeed, if we widen our theoretical scope to include our ideological, symbolic, and material relationships with the "other-than-human-life-world", we will develop new ways of "reading" violence. An eco-feminist approach is particularly relevant in the context of our current global ecological crisis and rapidly changing nature ideologies.

Chapter 2

Foregrounding patriarchy in a larger field of domination of nature

The oppression of nonhuman animals and nature is continuous with and linked to masculinist strategies of human domination. Understanding domination begs for the inclusion of human domination of nature as an operation of power. Patriarchy is frequently regarded as the basic hierarchy from which gender violence originates (Beechey, 1979). This chapter endeavors to make explicit those connections between patriarchy as a social domination that is deeply intertwined with the domination of nature. Hierarchy takes many forms—ethnic, gender, age, class, rural/urban, and so on. The common feature of all hierarchies, however, is dominance. Dominance can be defined as having power and influence over others. Treating difference as the basis for domination and superiority is the key mechanism of systems of command and obedience.

Patriarchy is one form of hierarchy, among many. In this chapter, I first give attention to patriarchy as its own apparatus and explore some of the features, critiques, and variations in patriarchal manifestations and their link to gender violence. I then go on to explore two key pillars of all hierarchical systems (including patriarchy): dominance and dualisms. I then move on to more firmly situate patriarchy within a larger field of human domination over nature by exploring "the hunt". The hunting of nonhuman animals fused a number of key elements: masculinized normalization of violence and a subjugation of the "Other" which is bound to the aggressive mastery and subjugation of nature. I continue to explore how these elements of domination that originated with "man the hunter and eventually warrior" produce familiar sexist scripts that can be linked to gender violence in a variety of ways. Ever mindful of the temptation to simplify power relations, I conclude with a section that considers the power relations explored here as more of a labyrinth than a ladder. In complicating these matrices of domination, I dismiss the possibility that achieving "gender equality" would remedy the problem of gender violence given the existence of other modes of domination, including human domination of nature.

Patriarchy and its many manifestations

The analysis of patriarchal societies has been instrumental in the study of gender violence and to a lesser extent in understanding environmental degradation. This chapter takes as a starting point that the answers we seek surrounding gender violence must include an understanding of a broader system of oppression of nature, inclusive of, but not exclusive to patriarchy. Since gender violence and environmental exploitation "have co-evolved with patriarchal capitalist-colonial power relations" (MacGregor, 2017), this chapter considers how ideologies of dominance in nature are similar to ideologies of dominance in gender violence. While patriarchy is often evoked by scholars to explain gender violence, it is important to consider in our theorizing that patriarchy is foregrounded by another hierarchy—human domination over "nature". Thus, gender violence stems from a logic of domination that is built on the domination of nature and the domination of the Other "as nature". Our situational, structural, and relational arrangements with the environment are connected to gender violence because power arrangements that characterize such relationships reflect our modern practice to place "man" over nature, as well as our yearning for control over nature.

In this text, I make extensive use of the concept of patriarchy as one key hierarchy deeply entangled with other forms of power arrangements in a larger field of domination. Patriarchy is of central importance here both as gender ideology and a central organizing feature in maintaining hierarchical social systems and social arrangements that reinforce domination in other areas of social life. Patriarchy, as it is employed here, references social contexts, not individuals. Since the 1970s, the tireless activism and research from scholars and organized feminist movements brought visibility to the issue of gender violence (more commonly referred to in this literature as "violence against women") and influenced huge leaps in social and legal policy (Schechter, 1982). The impressive literature on gender violence is copious, with volumes of work produced in the last four decades (Renzetti, Edleson, & Bergen, 2001; Yllo, 1993). The origins of the use of patriarchy as a theoretical concept to explain rape, in particular, can be traced to the pioneering work of Susan Brownmiller (1975).

However, the utility of the term "patriarchy" was called into question for being "undertheorized" (Kandiyoti, 1988). What followed was a chorus of contesting voices resulting in a widespread critique of the term "patriarchy". Subsequently, the tainted term was largely abandoned by scholars, but remnants of the idea could still be found packaged in alternative terms, such as "male-dominated society". Much of the gender violence literature maintains that explanations of violence should center on gendered social arrangements and power, but the configuration of language and concepts which convey this idea vary considerably (Dobash et al., 1992; Mooney, 2000; Yllo, 1993).

I present an alternative way to illuminate the problems of gender violence by employing patriarchy as a core theoretical concept, but with one caveat: this concept should be brought into conversation with ecofeminist scholarship. Gendered dominance is a complex feature of social life. The concept of patriarchy must be situated within other forms of hierarchy and domination from which it is inextricably embedded. The exploitation of the earth shares a common patriarchal core with other modes of domination. For example, the same dominance model that undergirds exploitation of the more-than-human-life-world parallels that of nations. The overdevelopment of wealthy nations exists in a colonial relationship to underdeveloped countries. The poverty in underdeveloped countries is a consequence of the exploitation and engineered dependence of rich nations. Maria Mies and Vandana Shiva (2014) point out how a similar colonial relationship exists today between "Man and Nature, between men and women, between urban and rural areas.... In order to maintain such relationships force and violence are always essential" (p. 56).

The term "patriarchy", as employed in this text, refers to "social arrangements that privilege males, where men as a group dominate women as a group, both structurally and ideologically; hierarchical arrangements which manifest in varieties across history and social space" (Hunnicutt, 2009). Patriarchal systems can be found at both macro and micro levels. At the macro level, the state, bureaucracies, religion, law, the market might be characterized as patriarchal. There is also the potential for patriarchal relations to exist at the micro level, such as in interpersonal interactions, organizations, families, or even patterned behavior between intimates (Gelles, 1993). Interpersonal dynamics exist in symbiotic relationship with social structure. Our individual experiences are nested within a macro-level gender order (Blumberg, 1984). Micro and macro patriarchal systems are co-constituted. There are both structural and ideological components to patriarchy where an organization might be patriarchal in structural terms, and individuals within the organization may (or may not) embrace patriarchal beliefs. Even as gender hierarchies are the central organizing feature of patriarchal systems, other modes of domination such as age, race, class, sexuality, religion, historical location, and nationality mediate gender statuses. These clusters of difference function to assign individuals varying amounts of power, social value, and privilege.

Patriarchal systems are ultimately rooted in separation from and domination of nature. In her sweeping exploration of the cultural origins of violence, Riane Eisler (1987) traces the origins of male-dominated social systems. These hierarchical and oppressive structures originate with human separation from nature, a precursor to hierarchy. Drawing on scholarship in archeology and history, Eisler (1987) showcases societies as far back as 2500 B.C.E. that were egalitarian and peaceful, or what Eisler calls a "partnership model" based on "linking rather than ranking". In contrast to the dominance model of social relations (patriarchy), partnership societies had less violence. Like Eisler (1987), Spretnak (1982) and Starhawk (1979) also trace the obliteration of agricultural, goddess-worshiping societies by marauding bands who worshiped the blade and a male god. The conceptual linkages among women and nature are evident in the shift from matrifocal to patriarchal societies. Murray Bookchin (2005), in a similar genealogical exploration, explores further the consequences of male supremacy, namely cruel, aggressive, and violent outcomes:

> Hence, patriarchal morality must bring her into complicity with the male's ever tremulous image of her inferiority. She must be taught to view her posture of renunciation, modesty, and obedience as the intrinsic attributes of her subjectivity, in short, her total negation as a personality. It is utterly impossible to understand why meaningless wars, male boastfulness, exaggerated political rituals, and a preposterous elaboration of civil institutions engulf so many different, even tribal societies without recognizing how much these phenomena are affirmations of male activity and expression of his "supremacy." ... But his increasing denigration of women and his transposition of otherness from a conciliatory to antagonistic relationship generates a hostile ambience in society—a meanness of spirit, a craving for recognition, an aggressive appetite, and a terrifying exaggeration of cruelty—that is to render man increasingly prone to the victimization of his own kind.
>
> (p. 194)

Patriarchies mirror a variety of other hierarchical arrangements. Erich Fromm (1955) articulated how the family reflects a similar patriarchal architecture of the patriarchal state and criminal justice system. As the family has changed over time, other institutions like religion and states have acquired a patriarchal authority that was once concentrated by male heads of family. The family may no longer serve as such a forceful cultural apparatus that supports other systems of domination, but most mainstream religions and the state are overwhelmingly still institutions of dominance that are secured with common patriarchal moral underpinnings. The Hebrew bible grants men a clear moral authority over women and children. Bookchin (2005, p. 191) elaborates on this patriarchal dominance and the familiar linkage of "woman" with "animal" to justify such oppression: "The bible provides ample evidence of the sovereignty enjoyed by the patriarch in dealing with his wives and children. To put it bluntly, they were his chattel, like the animals that made up his herds". Additionally, the criminal justice system operates with a model of punishment or "just desserts" that reflects patriarchal practices of control and domination of those who are weaker. These various hierarchical entities emulate old patriarchal family systems and share a common pattern: authority over others (Fromm, 1955).

Evidence for a connection between gender violence and various patriarchal arrangements abound. Domestic violence is more common in households characterized by allegiance to traditional gender roles (Walker, 1977–1978). Not only are the rates of domestic violence higher during wartime, but military personnel—individuals embedded in a deeply patriarchal, hyper-masculine institution—commit domestic violence at levels that exceed the general population (Nordstrom, 1999). If more extreme patriarchal ideology is connected to domestic violence, it is also true that men are not as account-able for victimizing women in societies with dramatic gender inequality. In societies where gender stratification is most extreme, authorities are perhaps less committed to preventing gender violence—a feature of overtly patriar-chal states (Blumberg, 1979).

If we consider the varieties of patriarchy (Hunnicutt, 2009), where men as situated in their own scheme of domination and acknowledge a range of different patriarchal manifestations among cultures, clans, and institu-tions, some of the more rigid manifestations of patriarchal organizations make violence more visible. Take militarized environments as a site where gender is an obvious product of social structural conditions once we con-sider how men are situated in their own scheme of domination within a culture that valorizes masculinized aggression. Feminist scholars in In-ternational Relations (IR) have written about the gendering of militarism. Enloe (2000, 2007) and Tickner (2001) have revealed how male dominance hierarchies characterize military institutions and explore the consequences of their hyper-masculinized cultures. Militarized entities, however, are also often charged with violent claiming and policing the natural world—both in our colonial history and maintenance of territories and borders today (Mohanty, Riley, & Pratt, 2008).

Military projects are so much about securing land, natural resources and conquest—controlling the earth's resources through aggressive, masculine might (Mohanty, Riley, & Pratt, 2008). Gloria Anzaldúa's poem "We Call Them Greasers" exposes rape as part of the violence of colonization. Point-ing to civilization and nature, Anzaldúa reveals the history of violent con-quest in American westward expansion into "virgin" territory where the white colonizer is associated with civilization and masculinity but also with violence (Coles & Zandy, 2007). Meanwhile, the colonized are represented as both feminine and close to nature, waiting to be possessed (Madsen, 2018). Annette Kolodny (1975) examined troping in exploration-narratives and found that generally the land is represented as female and the explorer is both male and a rapist. Both Kolodny (1975) and Anzaldúa deconstruct how the history of colonialization, militarism, conquering land, and dispos-session are rooted in gender violence.

That patriarchal belief systems are correlated with gender violence should be expected where restrictive attitudes about gender are embraced (Freeman, 2017). Freeman, writing in the Guardian, points out how both the military and Islam are male-dominated, male-focused, gendered hierarchical systems that practice strict sexual divisions of labor, and manipulate the social and sexual control of girls and women in service of their objectives. In patriarchal systems, a woman's "value" is determined by sexual purity, but also by race, class, age, appearance, and reproductive status.

While continuing to acknowledge that patriarchal configurations vary tremendously, and with an understanding that structure and agency, domination and contestation change and vary, one mode of "command and obedience" in some patriarchal systems (especially religious fundamentalist societies) is the perpetuation of the idea that the most important thing about a woman is her chastity (Skilbeck, 2001). Indeed, a complex matrix of coercion, agency, and survival can arise around this one issue of female sexual integrity. A whole system of values, protection, and freedom can swirl around sexual domination. In exploring gender violence in the Islamic State, Rajan (2018, p. 79) writes:

> Patriarchal ideologies, which privilege masculinity and marginalize women in most cultures globally, are further intensified in cultures/religions governed by principles of shame and honor. Those cultures more prevalent in nations in Global South and include Muslim-dominated geographies. Cultures guided by principles of shame and honor intensely relate masculine and social value, and hence, a man's social value/status, to the sexual propriety of the women in his family and, by extension, community. A woman may negatively impact the honor and reputation of the men in her natal and marital families and, also by extension, collectively the men in her community if she does not act modestly and guard her sexual purity. Even suspicions about her moral propriety (whether expressed by men or other women, and entirely unfounded) may dishonor and hence, devalue, a woman's sexual value, which in turn would dishonor her family. To mediate against the impact of her moral reputation on her family, her family (primarily the men), in turn, are forced to social devalue, distance themselves from her, and even ostracize her completely.

While Rajan's (2018) account reflects and extreme and rigid patriarchal manifestation, it is important to recognize the huge variation in how gender violence plays out. Thousands of dowry deaths—also called "bride burning"—are reported in India, Pakistan, and Bangladesh every year, but these rarely occur outside of these countries (Menon, 2003; Oldenburg, 2002). Rates of rape vary tremendously across time, culture, and space. Britain enjoys lower rape rates than in the USA (Messerschmidt, 1993) and some tribes have been shown by anthropologists to be rape-free (Sanday, 1981a, 1981b). While holding in our awareness this variation across time and space according to "degrees of patriarchy", it is possible to move away from "universalizing theorizing" (Beechey, 1979; Connell, 1990a; Smart, 1989).

Patriarchy, then, takes on multiple shapes within complex gender systems with varying forms of male domination. Flexible ways of viewing social arrangements have helped us to see systems of gender and dominance as fluid—more of a "flow than a structure" where patriarchal order, power, and customs are continually melting and reconstituting (Bauman, 2000; Carrington, 1994; Daems & Robert, 2007). The exploration of shifting historical situations reveals the instability of patriarchy over time, as well as how these systems are continually remade (Ryan, Walkowitz, & Newton, 1983). Thinking about flow, variety, and change in patriarchal systems as "varieties in movement" avoids essentializing and universalizing.

A key feature of patriarchy: domination

Murray Bookchin (2005), in *The Ecology of Freedom*, explores how humanity has unfolded within the limits of hierarchy and domination. Bookchin (2005) traces the landscape of hierarchy from our ancient past to the present, exposing how the curse of domination has created conditions of social conflict and subjugation for both humans and nature. In reviewing anthropological evidence from preliterate societies, or what Bookchin (2005) calls "organic societies", these social units could be characterized as sharing a strong internal unity, where their egalitarian outlook stems from a unified structure rooted in kinship ties, and a harmonious sexual division of labor and age groups. It is in the decline of organic society where Bookchin (2005) locates the rise of hierarchy, the result of which has created a system "whose appetite for rule is utterly unrestrained" (p. 449).

Bookchin (2005) stresses that organic societies had an orientation that was ecological. This ecological orientation in organic societies influenced values and behavior toward their communities that was distinctly egalitarian. It will be hard, perhaps, for the reader who has only ever known hierarchical arrangements to conceive of such communal communities. Bookchin (2005) maintains that these early ecological societies "knows no 'king of beasts' or 'lowly creatures'—such terms come from our own hierarchical mentality" (p. 69). Indeed, in organic societies, independence, complementary roles, and a unity of difference or a unity of diversity prevailed. This egalitarian outlook extended to each other, but also to nature—they were within nature, rather than above it or below it. Bookchin (2005) claims that rather than seeing themselves as "lords of creation", humans in organic communities regarded themselves as interdependent with nature and occupied complementary roles.

As hierarchical social arrangements replaced egalitarian ones, this internal unity began to break down and forms of stratification and differentiation gave increasing authority to males and elders (Bookchin, 2005). Due to male's changing roles with the expansion of the civil sphere, male supremacy over women began to surface (in the next section, I will expand more on the eminence give to men due to the "hunt"). This hierarchical mentality that eventually took hold "ranked the most miniscule phenomenon into mutually antagonistic pyramids erected around notions of 'inferior' and 'superior'" (Bookchin, 2005, p. 72). It was here that human animals then adopted the entitlement to dominate. Members of egalitarian communities are still characterized by difference. It is possible to have difference without domination, of course, where relationships are not based on command and obedience. Bookchin (2005) notes that even the most egalitarian societies are not homogenous. Instead, members may occupy roles that define their responsibilities, but need not be structured by a hierarchy, nor involve inferiority or obedience. What is key here is that this egalitarian orientation extended beyond the human.

More contemporary manifestations of dominance over nature are bound up with science and industrialization. Modern science and industrial capitalism identify with a masculine norm. Merchant (1980) details a history of the scientific revolution whose progress and innovation rest on violence and exploitation. Carolyn Merchant's (1980) *The Death of Nature: Women, Ecology and the Scientific Revolution* is regarded as one of the founding ecofeminist texts in the West. Merchant explores the connections between environmental crisis, capitalism, colonialism, science, religion, philosophy, and patriarchy. Merchant (1980) traces the ways in which subordination of women was "naturalized" by shifting cultural values, which parallel the shift from reverence to domination of the earth. Carolyn Merchant argued that the control of nature was closely bound to men's control of women— an ideology of dual dominance. Merchant claimed that modern capitalism was achieved by the domination of nature, which is feminized, and which only reinforced the subordination of women. Violence comes about in the maintenance and construction of social and economic hierarchies. Indeed, our current social, political, and economic configuration is designed to continue to extend the violent commodification and exploitation of animals and nature. Just as the exploitation of a feminized natural world was central to the rise of capitalism, technology, and modern science (Thompson & MacGregor, 2017), the subjugation of a "gendered" earth has reinforced the subordination of the Other.

Merchant (1980) demonstrated that it was the enormous power over nature facilitated by modern science that allowed the West to enjoy success in colonial projects. Dominant groups and ideologies continue to legitimate their rule to this day. In this contemporary moment, it is modern Western societies (the most dominant) that bear the greatest responsibility for environmental destruction. Androcentrism (male dominance) and anthropocentrism (human-centeredness) are the root cause of environmental degradation and alienation from the more-than-human-life-world, where the environment is conceptualized as a set of resources for human consumption (Cudworth, 2005).

The notion of domination is a quite complex system of social relations that is both a "descriptor for relations of power" and is fundamentally about limiting life chances (Cudworth, 2005). Cudworth's (2005) conception of domination includes three levels: marginalization, exploitation, and oppression. This differentiation is important because, as Cudworth (2005) argues, different aspects of the environment are differentially dominated. While the entirety of the environment can be seen to be subject to human control—some aspects are exploited while some are oppressed. For Cudworth (2005), gendered domination is intertwined in a multiplicity of systemic domination based on various forms of difference. If we look at the dynamics of various systems of social power, they look similar, which is why a multiple systems approach is preferable. "By tracing the strains of masculine domination through culture, nature, body, labour, logic, technology, culture etc... ecofeminism shows how all social movements share a common denominator" (Salleh, 1997, p. 36).

Systems of domination share a number of features. They find expression through cultural imperialism, violence, exploitation (of resources, labor, sexuality), marginalization (social exclusion), and powerlessness (lack of personhood, meaningful rights, efficacy) (Young, 1990, pp. 56–64). The amount, extent, type, and degree of oppression will vary depending on the extent of overlapping identities in marginal groups, as well as the relation of our membership in groups to institutions and practices of social domination (Young, 1990). Systems of domination all feature a value-laden hierarchical order where privilege, power, and superiority are attributed to "Up", while inferiority is "Down" (Warren & Cady, 1994). Power is exercised "over" rather than "within" and there is a logic of domination in play that reproduces and maintains certain subordinations. And finally, systems of domination depend on dualisms where higher status is given to one component of an oppositional pair, such as man/woman or reason/emotion (Warren & Cady, 1996).

Systems of domination can operate in very subtle ways. Antonio Gramsci (1992) argued that hegemony curtails what is possible for us, as it draws the confining maps by which we live, and in this sense, hegemony is a mechanism of psychological control. And in these subtle psychological operations of hegemony, we are conditioned to believe that the exploitation of people, animals, and the earth is both inevitable and natural. Both gender violence and domination of the earth operate with these ideological mechanisms of domination that work to maintain the status quo. This sort of ideological manipulation may lead us to believe that certain groups of socially marginal people and nature are less worthy, thus paving the way for oppression and violence to play out. Finally, domination within hierarchical systems is fed with dyadic frameworks which rank, sort, and assign value to different life forms and social groups.

A key feature of hierarchy: dualisms

The very foundations of Western thought serve to naturalize male domination through dualistic imaginings. The underlying ideology of subordination is gendered dualisms, which are political, exclusionary, hierarchical blueprints. These core oppositional dualisms that organize Western thought include mind/body, culture/nature, human/animal, reason/emotion, public/private, production/reproduction, and so on. Feminist analysis exposed these hierarchically ordered categories as gendered with the term on the left side being associated with masculinity and femininity on the right side, which seemed to uphold a patriarchal ideology (Haraway, 1991; Mathews, 2017). For instance, reason/emotion, human/animal, man/woman, culture/nature, and mind/body are dualisms in which the first construction is regarded as superior to the second construction.

While a number of scholars have written on this subject, Val Plumwood (1993), in particular, showed how dualisms serve as agendas of domination in a whole host of regimes, not just gender, but also class, race, ability, species, etc. Plumwood (1993) demonstrated how dualisms are vehicles to rationalize the subordination of many groups, as well as nature. In any dualistic construction, there is the potential to create an inferior "other" and uphold various forms of privilege. These binary frames contribute to various injustices and distort our relationship with the more-than-human-life world. The privileging of reason, for example, is deeply entangled with histories of racism, sexism, patriarchy, imperialism, classism, and speciesism.

The domination of nature extends to political domination. The use of dualisms such as human/animal and culture/nature plays out in a national context where the ground of difference is constantly (re)produced between "us" and "others". The "other" in these narratives is cast as a danger to "our" way of life. Speaking about animals and nature as "others", the difference between "us" and "them" is often pronounced in terms of the violence "they" can initiate against "us". The fear of nature also organizes our lives of separation and threat. If a latent danger always lurks in nature, we must separate from it, suppress it, harness it, and restrict our movement to developed spaces. Such assumptions shape our understanding of the environment as requiring violence, as a violent relational field which then necessitates our continued violent response against its "threat".

Ontological dualisms exist to afford privileges to certain social groups and to devalue Others so that they can be justifiably exploited. These binary relations exist in a value hierarchy and structure our social institutions, ideologies, and laws, largely determining who can violate and who can be violated (Kheel, 2008). For Val Plumwood (1993), the key currency in legitimating anthropocentric authority—and by extension patriarchal authority—is "rationalism"—the philosophical tradition that elevates humans over all others based on their ability to reason. This insight from

Plumwood highlights the two components to domination—ideological and material. The ideological draws parallels in ways of thinking and believing while the material dimensions involve the manifestation of environmental destruction and violence among people and actual spaces. This dualistic technique is employed in such a way where reason is privileged and therefore anything close to nature will be considered inferior.

Bailey (2005) notes that reason has historically been the measure of humanity while anything that comes from the body, such as emotions, feelings, and alternative ways of knowing, is inferior. From this key assumption, racist and gendered belief systems began to take root. But Bailey (2005) goes on to say that reason came into favor precisely because of oppression. This insistence that reason is associated with full personhood, as an index of worth, has historically served as de facto justification for extending or withholding compassion. A key part of both anti-violence and earth justice work, then, includes dismantling powerful cultural stereotypes and rejecting false dichotomies of female/male, private/public, and victim/perpetrator binaries. I now pivot to explore how the "hunt" has been instrumental in the development of patriarchal arrangements, the fusion of violence with a masculinist discourse and the ways in with the hunt has informed gender violence.

Hunting in patriarchal societies and gender violence

In hierarchical societies, men and masculinity are associated with life-taking activities such as hunting, war, and violence, while women are associated with life-giving activities of reproduction, childbirth and care for the young. It makes sense then, that "life-taking" activities would become points of gender polarization and historically, a major mover of cultural practices. In *The Second Sex*, Simone de Beauvoir (1968) famously said that in "Eurocentric tradition, it is not giving life but risking like that raises man above animal". Beauvoir (1968) conveys the values afforded to "life-taking" activities above "life-giving" practices. Hunting is often portrayed as a universally normative aspect of human–animal relations. Beauvoir's quote exposes the patriarchal foundations of violence against animals. Scholars who followed Beauvoir (1968) exposed the gender/race/class specificity of a violent practice that is often portrayed as a "natural" component of human–animal relations.

The practice of humans hunting nonhuman animals has served to heighten masculine temperaments, exaggerating aggressive and violent aspects of "manliness" (Bookchin, 2005). We can see evidence for the tendency for hunting to magnify violent characteristics among males by comparing horticultural societies with hunting communities. In largely food-gathering and horticultural societies, there is evidence for a temperament that regarded women and men in complementary ways, while in pastoral and hunting societies, there is more of a tendency for social norms to favor men (Bookchin, 2005). Because males in hunting communities are specialists in violence, masculine traits become associated with strength, courage, and physical feats (Bookchin, 2005). Concomitantly, female traits decline in value, and are seen as less-than-human, since live-giving traits are shared so closely with the natural world.

This gender polarization around hunting led to the domination of women and Othered people in many aspects of social life. Collard and Contrucci (1989) traced the historical link between hunting, or the incorporation of meat into our diet, and the domination of women. As our early human development reveals, hunting was often framed in terms of the gendered division of labor. Hunting, by all accounts, is predominantly practiced by males (Kheel, 1995). The practice of hunting rendered men active, women passive and created gendered social arrangements where women became dependent on men for food and protection. That hunting is a fundamentally violent practice is important here precisely because this dynamic of dependence on a gender-segregated violent activity eventually became naturalized (Collard & Contrucci, 1989). As the belief that killing animals for food was necessary, hunting became, according to Collard and Contrucci (1989), a mode of destruction that is rationalized as a "noble purpose".

While the hunt solidified a male-dominated culture with profound inequities to follow, this practice has lent a sort of legitimacy and normalization to violence of all kinds. While killing would give rise to feelings of revulsion in most people, the narrative of the noble hunt allows this type of violence to pass by unrecognized. Emel (1995) reflects on the ability for hunters to admire the thing they have murdered as requiring a "curious detachment" and highlights how disturbing it is to shut off feelings in order to facilitate death (p. 724). Collard and Contrucci (1989) contend that as long hunting is regarded as an evolutionary leap, the practice will continue to be "naturalized", and human animals are then relieved of responsibility for the killing and brutality which continues under the guise of a "noble purpose" (p. 33). This sets up a situation of identifying a "bad" violence to justify a "good" violence. The noble purpose or the "civilized atrocities" committed against those considered "savage" has a haunting similarity to totalitarianism and extermination of human populations (see Rogin, 1987).

In this sense, violence is rationalized because of the false belief that such brutality advanced our species and our civilization. Indeed, there are links between "hunter" and other violent roles occupied by men. Bookchin (2005) explores how this development toward a male-oriented culture introduced elements of domination where the tribe tended to "assimilate his temperament as hunter, a guardian, and eventually as a warrior" (p. 148). Later during colonialism and westward expansion, the hunt was symbolic for not just wilderness, but a sort of wildness—an enemy to be conquered. This "slaying the beast" reflects a colonial mentality where animals, nature, and dark-skinned people were to be conquered (Emel, 1995). Mackenzie (1987) elaborates: "The slaying of the dragon represented the ultimate in justified killing which, when transferred to big-game hunting, represented the victory of civilized man over the darker, primeval and untamed forces still at work in the world" (p. 81).

The more differentiated we become with nature, the more the internal unity of our communities attenuate. Nibert (2013) traces the collapse of egalitarian, communal societies due to the introduction of the highly gendered practice of organized hunting of nonhuman animals as well as the continued legacy of violence. Drawing on existing anthropological research, Nibert (2013) reports that organized hunting likely began 90,000 years ago, a conservative estimate. In all likelihood, the onset was much later. At this time, men's role in hunting cultures began to be associated with this violent activity. Through the repetitive practice of stalking and killing of nonhuman animals, the propensity among human hunters to exact violence against humans increased (Nibert, 2013). As men became increasingly concentrated in hunting roles, women became increasingly concentrated in domestic and childcare roles. Concomitantly, the status of women declined while communities began to grant eminence to males who displayed the traits of a "good hunter". In other words, as men are prized for becoming specialists in violence, hierarchy increasingly intrudes into domestic life, resulting in the subjugation of the Other. Nibert (2013) laments the decline of women's status to that of property as being bound up with the domestication of nonhuman animals:

> As in the stratified social systems of other nomadic pastoral peoples, the role of women in Africa was undermined by the practice of exploiting large numbers of domesticated animals. In African nomadic pastoralist societies—which, like other pastoral societies, were patriarchal, hierarchical, and stratified—a man's status was determined by the number of captive animals he controlled. Male responsibility for and control of domesticated animals developed in part out of a need to defend against their abduction by other pastoralists; furthermore, other animals—especially cows—increasingly came to represent bride wealth. Bride wealth is a payment from the groom's family (or, more rarely, the groom himself) to a girl or woman's family for the right to marry her. In exchange for the payment, the male obtained various rights over the "bride," and children resulting from the arrangement became his "property."
>
> (p. 156)

Animal husbandry began roughly 11,000 years ago and ushered in a speciesist model where animals became slaves (Nibert, 2013). Removed from their natural habitat, these nonhuman animals became dependent on humans for food and protection. This shift in power arrangements coincided with religions that embraced hierarchical beliefs, in particular, worshiping a sky-god (Griffin, 1978). In contrast, Bookchin (2005) contends that horticultural communities, which developed some 10,000 years ago, were able to separate themselves to a degree from their dependence on the hunt. As a result, there was an accompanying social shift of "male the hunter" imagery to more peaceful pursuits.

Remains from Neolithic communities provide evidence of peaceful societies (Bookchin, 2005). These early horticultural societies left symbols indicating very little war or inequality. Instead, their artifacts reflect communion with nature and other humans, an internal solidarity. And while there were weapons, the emphasis of life seemed largely focused on nature's abundance and peaceful lives close to the hearth (Bookchin, 2005). Moreover, anthropological evidence suggests that women and men enjoyed a high degree of parity within these societies and the distinctions between "public" and "private" or "home" and "world" did not exist in organic societies (Bookchin, 2005).

The belief that hunting has formed the basis of our social life and success of our species is a frequently exaggerated claim and has led this form of violence toward animals to be taken for granted. Indeed, anthropologists have shown that historically human survival relied much more heavily on skills needed for the gathering and consumption of vegetables, rather than killing animals for food (Reiter, 1975; Sapontzis, 2004). In recent history, hunting has been regarded as a normative practice, yet support for hunting has been in decline since the 1970s (Herzog, Rowan, & Kassow, 2001; Irwin, 2001). Indeed, a majority of Americans indicate that they oppose hunting (Irwin, 2001). The decline in both the practice of and public support for hunting can be attributed to the rise of the animal rights movement, the decline of leisure time and less available land where hunting is permitted.

But even as the practice of hunting nonhuman animals has declined, the familiar sexist scripts—rooted in a devaluing of nature—have been imported from the hunt into similar sexist logics repeatedly used to legitimize militarism and violence (Leatherman, 2011). Rape has historically been wielded as a weapon of war, while "protection" of particular females associated with race and nation may serve as a justification for war. There has been much discussion of rape as a weapon of war and how "protection" is a gendered notion (Leatherman, 2011). For example, war seeks legitimacy by characterizing the state as the protector where men are fighting to defend women (e.g. protecting family) at home or abroad (Leatherman, 2011). Pattinson (2008) calls this the "masculinization of the protector role" and the "feminization of the one who needs protection". The morality theater where the state plays the role of the savior requires a helpless, forlorn, female subject in order for the narrative to be successful. So much of this ideology stems from our differentiating "man" from nature and positioning "woman" closer to the earth. Bookchin (2005, p. 193) writes, "in a civilization that devalues nature, she is the 'image of nature', the 'weaker and smaller', and the differences imposed by nature between the sexes become the most humiliating that can exist in a male-dominated society ... a key stimulus to aggression".

Gender ideology that includes elements of domination where men are "hunter, a guardian, and eventually warrior" (Bookchin, 2005) is leveraged in modern day violent enterprises. Consider military projects that involve the state casting women as vulnerable subjects who need to be liberated with warfare. Prior to the 1990s, it was rare to see policymakers prioritizing gender violence as a social problem. Even in human rights discourse, political actors rarely mentioned rape as a human rights problem prior to 1990 (Harrington, 2016). The 1990s, however, were marked by a sea change of political interest in violence against women, particularly as a national security issue (Enloe, 2000; Harrington, 2016). After the 9/11 terrorist attacks in 2001, U.S. security issues have centered largely on terrorism. Employing a gendered ideology has worked to shore up public support and justification for continued warfare. By focusing on "vulnerable" women in "war on terror" militarized spaces who ostensibly need rescuing, provides a justification for violent action by the state (Russo, 2006).

The controlled violence of hunting is a ritual that demonstrates a relationship that men have to nature, where masculinized violence is reinforced through symbolic acts of domination over nature. But dominance through militarism is another key iconic display involving archetypal meanings illuminating how the practice of gender violence is sustained through symbolic displays of the "man triumphs over nature" motif (Phillips & Rumens, 2016). The state itself is a gendered institution, and it engages in the masculinized "rescue" of females facing perilous circumstances (by other males). This power imbalance fueled by beliefs about female vulnerability enables the state to claim an image of paternal benevolence (Hollander, 2001; Leatherman, 2011). This terrain becomes confusing when state actors evoke a vocabulary of liberation. The reality behind such emancipation rhetoric may be a set of imperialist interest, often related to development, security, and economic agendas (Leatherman, 2011; Mason, 2017).

The practice of hunting can be linked to gender violence in other ways. Carol Adams (1994) and Marti Kheel (1995) are perhaps most famous for arguing that hunting is a form of violence that informs other expressions of male domination and oppression. Collard and Contrucci (1989) contend that the male-dominated practice of hunting in some parts of the world may have led to a set of social practices that were informed by the hunt, namely destroying nature, violence in war, raping women, and enslaving other human animals. Collard and Contrucci (1989) contents that the hunt informed a culture of brutality that was instrumental in shaping our patriarchal social order. When hunting is considered sport, as it is in contemporary times, there is a fundamental glorification of violence that provides a "paradigm of legitimacy" for other types of violence, including rape.

Ascione (2001) contends that committing animal abuse in childhood may impair the healthy development of empathy. It follows then that the killing involved with hunting might also lead to an impaired sensitivity toward human–animals. It is a notable pattern that most hunters are male, and that male socialization focuses on dominance and aggression while minimizing empathy (Coltrane, 1998). Gender violence and violence against animals derive from a similar frame of mind—one in which subjugated beings are regarded as entities to be controlled, managed, and manipulated—often with an aim to distinguish masculine identity (Kheel, 1995). The decline in women's status as males became increasingly exalted for becoming skilled in violence, along with the fusion of women with domesticated animals, reveals how our material relationship to the nonhuman lifeworld structures relations of power. These observations challenge us to closely examine our assumptions about the naturalness of violence, particularly in its gendered aspects.

Situating gender violence within a labyrinth of power and matrix of domination

Since the concept of patriarchy has been the subject of extensive debate, it is important to acknowledge some of these critiques. The ways in which patriarchy has been deployed to explain gender violence has been criticized for simplifying power relations, implying a "false universalism", ignoring differences among men by casting men instead as singular group, and for being limited in its explanatory power (Merry, 2009; Price, 2005). In this section, I strive to complicate and expand our conception of power relations under patriarchal systems by situating gender violence within a larger matrix of domination and conceiving of power as more of a labyrinth than a ladder. First, however, I address the limitations of trying to reverse patriarchy and remedy the problem of gender violence by striving toward "gender equality".

When looking to ameliorate the problem of gender violence, scholars often identify gender inequality as the root cause of the problem. Numerous empirical studies explore the relationship between gender inequality and rates of violence against women (Austin & Young, 2000; Avakame, 1998, 1999; Bailey, 1999; Bailey & Peterson, 1995; Baron & Straus, 1987; Brewer & Smith, 1995; DeWees & Parker, 2003; Dugan, Nagin, & Rosenfeld, 1999; Ellis & Beattie, 1983; Gartner, Baker, & Pampel, 1990; Levinson, 1989; Pridemore & Freilich, 2005; Smith & Brewer, 1995; Stout, 1992; Titterington, 2006; Vieraitis & Williams, 2002; Vieraitis, Britto, & Kovandzic, 2007; Whaley, 2001; Whaley & Messner, 2002; Yllo, 1983; Yllo & Strauss, 1984). The results of this work, taken as a whole, show that there is not a direct relationship between gender equality and violence against women, that the results are very mixed.

The promise that achieving economic equality among women and men would reduce gender violence has been met with skepticism (MacKinnon, 1983). As long as other systems of domination prevail, gender equality alone would not lead to a nonviolent society. Murray Bookchin (2005) writes:

> There is no reason to believe that a gender integrated police force—or for that matter a gender-integrated army, state bureaucracy, or corporate board of directors (given the very nature of these institutions as inherently hierarchical) would lead to a rational and ecological society.
> (p. 27)

Marxist feminists have pointed out that as long as capitalism exists, sexist beliefs would remain alive and well independent of gains in greater gender equality (Jagger, 1983). Gender violence plays out in a sweeping matrix of domination (Collins, 1991) characterized by both structural and ideological features.

Gender inequality is a very important issue but it is doubtful that parity in income, rights, privileges, and political representation would reverse patriarchal social systems, given the existence of other systems of domination. Because patriarchal systems are bound up with other systems of domination, they must be understood within extended fields of hierarchy where whites dominate people of color, where old dominate young, where developed nations dominate developing nations, humans dominate nature, women dominate women, men dominate women, and men dominate men.

African American women are at a higher risk of rape (Messerschmidt, 1993) which reflects a racialized hierarchy. Rape survivors are more likely than not to be economically marginal, earning less than $10,000 a year (U.S. Department of Justice, 1996), which reveals that domination based on socioeconomic status coexists with patriarchy and racial domination. Teenagers and young adults experience some of the highest rates of rape, exposing another hierarchy based on age (Schwendniger & Schwendinger, 1993). Mutually constituted hierarchical systems of race, sexuality, ability, gender, age, class, and so on, are aligned with patriarchal systems. Moreover, both men and women are complicit in sustaining and reproducing these systems (Dinnerstein, 1976). Because any one of us might enjoy one position of domination, afforded by our Whiteness, our age, or social class, we might find an alliance with patriarchal interests, regardless of our gender. There are abundant examples where women enjoy great respect and status in their patriarchal family systems as they age (Kibria, 1990). This hierarchical reward may result in a motivation to maintain their particular patriarchal ideology (Kibria, 1990). At any given time, any individual occupies multiple positions of power, privilege, and disenfranchisement in a landscape of domination. This means that in our endless manifestations of unique social contexts "an individual may be an oppressor, a member of an oppressed group, or simultaneously oppressor and oppressed" (Collins, 1991, p. 225).

Systems of command and obedience, or relations of domination, are enormously complex. Hierarchy does not play out in a unidirectional top-down fashion but is instead a "terrain of power" (Flax, 1993). In such terrains, individuals possess ever-shifting amounts and types of power. Even the most disenfranchised individual is engaged in some power relations. In hierarchical systems, there are, in fact, multiple "sites of power". Scholarship on peasant resistance, most notably, *Weapons of the Weak* (Scott, 1985), explores strategies of resistance that peasants mobilize against their continued oppression, revealing that subordination gives rise to innovations in resistance and self-protection mechanisms. These might include forming alliances with others. For example, recently resettled Vietnamese women refugees found protection from abuse spouses by forging informal networks with other immigrants (Kibria, 1990). When women commit violence, this behavior can also be connected to systems of domination, submission, and resistance. When women kill abusive partners in self-defense, they are reacting to this system of oppression (Browne & Williams, 1989). When unmarried, pregnant Hindu mothers in Fiji commit suicide rather than face stigma, ostracization, and humiliation, they have internalized patriarchal oppression (Adrinkrah, 2001).

As long as stratified social systems attach power, value, and rank to particular social locations, opportunities for resistance and protection will be available to individuals based on their varying social positions. Class confers power upon subordinated women while subordinated positionalities generate opportunities for creative resistance strategies and protective mechanisms. Any individual possesses differing degrees of power depending upon our differentiation in this "matrix of domination" (Collins, 1991). Power and privilege are fluid, never achieved "once and for all", but instead shift along with movement in the life course through age, education, marital status, geographic location, and so on.

One paradoxical feature of patriarchal systems, and a complication in the study of gender violence, is the patriarchal (paternalistic) value of protection of women. Patriarchal ideologies specify that some women, but not others, should not be harmed. In this sense, the "protection" of women by men ultimately serves as a social control mechanism and an instrument of repression. The masculinized protection of females is a challenging subject to explore because it is so closely aligned with anti-feminist ideologies (Chafetz & Dworkin, 1987). But it is important to explore how "Othered" individuals may be simultaneously subject to the aggression and mercy of the more powerful party. In this sense, acquiescing to "protection" reifies vulnerability and powerlessness and existing hierarchies. And yet, paradoxically, the real threat of violence may require accepting the protection offered up in patriarchal systems. Patriarchal systems are engineered to keep the Other in a state of dependence, precluding full emancipation.

A microcosm of the varying degrees of risk for, and protection from, gender violence can be observed in Jody Miller's (2001) ethnographic work on gender and gangs. In the gangs that Miller (2001) studied, girls were sexually exploited but were also protected from the deadliest gang activity. The patriarchal practice of "protecting" certain women is bound up with their perceived "worth". The practice of Purdah in some Muslim cultures sequesters women from the sight by restricting women's mobility and clothing. Purdah confers a certain status upon women as a "protected group" but at the same time, the conditions of such protection are that these women are denied full, independent participation in life (Cain, Khanam, & Nahar, 1979). Since the "protection of women" serves as a social control mechanism, women who violate gender-specific normative standards may have their protections rescinded. The complexity of patriarchal relations—systems of command and obedience—means that Othered individuals experience varying "degrees of risk and protection".

Conclusion

How does ideology extend from relations of domination in one sphere (human–human) to another (human–earth)? By offering a general map of interaction that extends to various relational configurations. Ideology constitutes beliefs but also sets of social "scripts" that we use to navigate our social world. Our worldview, which is shaped and inherited within a content of dominating the earth, provides us with basic parameters for the operation of life. We then continually reproduce our ideology through daily performances, enactment of our beliefs, and institutional practices. Given that humans largely live in a context where the extreme exploitation of the earth is widely practiced, humans are therefore largely guided by an ideology that endorses this domination—a belief in the superiority of human animals and a belief in an entitlement to take from the earth anything we desire. This widespread belief in human superiority and entitlement is a very gendered phenomenon, closely tracking a patriarchal worldview. This ideology of superiority forms the basis of our understanding of the world which impacts our human interaction.

The hunt is a mode of violence that is imported into other modes of violent domination. Given a long history where men were exalted for becoming highly skilled in violence against animals, hierarchy defined by gender and might increasingly informed other areas of life, resulting in the subjugation of the Other. Moreover, the masculinized ritual of repetitively stalking and killing of nonhuman animals informed a wider culture of violence against humans (Nibert, 2013). This violent domination of the earth and the belief that its nonhuman inhabitants are violable supports a range of other social practices, including interpersonal violence. While the hunt solidified a male-dominated culture with profound inequities to follow, this practice has lent a sort of legitimacy and normalization to violence of all kinds. Through practices such as the hunt, the more differentiated we become with nature, making egalitarian, communal societies less possible. The introduction of the highly gendered practice of organized hunting of nonhuman animals has informed a continued legacy of violence.

The larger infrastructure of these violent practices included patriarchy, but patriarchy is foregrounded by another hierarchy—human domination over "nature". Thus, gender violence stems from a logic of domination that is built on the domination of nature and the domination of the Other "as nature". The amelioration of gender violence, animal subjugation, and environmental degradation require the construction of more just societies. These co-occurring issues have similar root causes; both are outgrowths of the same system of hegemonic structures of exploitation. The domination of the earth is yoked to violence. This relationship is gendered to the extent that sexism is bound up with other forms of domination and because nature is gendered "female".

Chapter 3

The intersection of masculinity, gender violence, and domination of nonhuman animals

Introduction

In this chapter, I examine the congruence of hegemonic masculinity and violence toward nonhuman animals as an impetus to understand gendered violence. Deep-seated notions about aggressive masculinities have shaped relationships with nonhuman species and given rise to other complex forms of oppression. Standards for "manhood" unquestionably shift across time, culture, and space (Kimmel & Messner, 2004). While it is not possible to make any universal claims about desirable forms of masculinity, we can detect themes of virility and prowess in various forms of violent masculinities and even the construction of a masculine identity through killing.

I rely heavily here, and expand on, the work of Carol Adams (1990, 2003, 2010, 2011) who exposed the ways in which violence against gendered and racialized human Others and nonhuman animals are linked, and Marti Kheel (2008) who suggested that masculine identity is a chief element in violence against nature. Through analytical concepts of the "feminized animal" and the "animalized woman", Adams (2011) considered the gendered dimensions of both reproduction and consumption in nonhuman animal food production. In this chapter, I hope to show how the violent manipulation of nonhuman animals' bodies is deeply gendered, sexualized, and racialized (Twine, 2014). Kheel (2008) considers how the achievement of manhood, as an ideal, involves a violent transcendence of nature—a separation of self from the natural world and idealizing rising above the natural realm. This analysis proceeds with an understanding that identification and representation are instrumental to oppression. If some constructions of masculinity embody identifications of masters and hunters, this will shape some behaviors, practices, and expressions.

These themes explored here are initially foregrounded by a consideration of hegemonic masculinity as a master discourse. There is no such thing as a singular masculine identity, but rather an infinite number of configurations of gendered performance. Despite these variations, there is a hegemonic ideal which has tremendous cultural purchase (Kimmel, 2006). This ideal construction of manliness includes a sharp distinction from "female", which includes feminized traits, such as care and empathy and associations with the natural world. The distancing from the "female imagined world" may be accomplished in a variety of ways. A "heroic" and violent subjugation of the nonhuman is one mode of "conquest" that achieves this separation (Keller, 1986). Despite the multiplicity and fluidity of actual gender expressions, this binary, value-laden hierarchical construction still exists in our cultural imagination and social organization. This hegemonic ideal is not only formed in opposition to women, but also to subordinated masculinities and operates as a "diffuse worldview that inheres in institutions, power relations, and ideas. The significance of this is that people may hold masculinist ideas independently of their conscious awareness" (Kheel, 2008).

In this chapter, I take a rather circuitous route through the exploration of masculine narratives involving the violent subjugation of animals and gender violence. I first consider hegemonic masculinity at a master discourse. I cannot emphasize enough that my reference to masculinity does not involve a unitary subject nor a generalization about all men, but rather is about expressions of masculine ideals in Western patriarchal traditions. I then move on to consider hegemonic masculinity as anti-ecological. To the extent that a masculine ideal rests on separation from nature, transcendence of nature, autonomy, risk, struggle, and distancing from the female, nature becomes the "Other" which must be subdued, often violently.

The expression of this superiority over nature may take a variety of forms but is most evident in domination of nonhuman animals. I then move on to explore a variety of ways in which this violent performance of dominance of the nonhuman plays out. I first consider the eating of animals as one process by which male dominance is expressed. I then move on to blood sports, including rodeo, where the bloody spectacle of controlling and manipulating nonhuman animals in competition and conquest is linked to the compulsion to prove a dominance over nature. I explore the ways in which blood sports have become racialized, where cruelty toward animals falls along the lines of acceptability based on race. I then leap to the fusion—often eroticized of females and animals and explore the ways that violence is entangled with both gender and animality. In examining the relationship between violence and the subjugation of human and nonhuman animal bodies, I conclude by making the case that the nonhuman animals must be taken up as a subject of serious study if we are to understand gender violence, and related oppressions, given that these two issues are intricately entangled.

Hegemonic masculinity as master discourse

Hegemonic masculinity has been explored in both the gender violence literature and environmental studies scholarship (Connell, 1990a). The environmental movement challenged the exalted location of the white, straight, dominating male (Connell, 1990a). Critiques of Euro-Western cultural constructions of masculinity include male identities being characterized by competition, achievement, hunting animals, dominance, strength, conquest, and constructions of maleness are usually formed in opposition to femininity (Gaard, 2014, 2017). It is important to stress that individual men are not at issue here, but rather what Val Plumwood (1993) framed as a "master" discourse. This "master" discourse is not an actual embodied experience of being male, but rather a hegemonic masculinity—an exaggerated, normative, idealized cult of virility. That said, we may see some of these inflated masculine characteristics reflected in some individual men. But we are more likely to detect them in cultural themes, societal characteristics, and institutional features. While this analysis does not reference individual men, this "master discourse" does, however, impact our lived experience.

Social systems characterized by male domination allow men to demonstrate that they are different from, and better than, women and socialize individual men with hierarchical aspirations. Chodorow (1978) argues that masculine identities are secured through the expression of prestige and authority. The pursuit of prestige among males includes an incentive to reject all things feminine. This system ends up marginalizing not only some women but also other men who do not reflect normative standards. That men are situated in their own matrix of oppression means that socially marginal, racialized, or "Othered" men may experience "blocked access" to prestige generating opportunities.

It is also true that the "master" discourse surrounding masculinity has been shifting as historical changes have challenged hetero-patriarchal power structures. It is common for scholars, public intellectuals, and clinicians to describe masculinity as being in "crisis" (Kimmel, 2015). The contemporary crises in masculinity are commonly traced back to "disruptions" caused by key social movements from the 1960s and 1970s, namely civil rights, gay rights, antiwar, and women's rights movements. Each of these movements disrupted the idea that a dominant male identity (i.e. heterosexual, economically successful, socially valued, dominant and dominating) was supreme (Faludi, 1999).

One such disruption in masculinity is the destabilization of the hegemonic white male. Kimmel (2015) examines the phenomenon of "angry white men" as stemming from a loss of entitlement promised by traditional gender ideologies. Messner (2016) points to white men's increasingly restricted access to power, women, and privilege in racial hierarchies, while working-class men may experience relative deprivation downward as they perceive a loss of opportunities to minorities and women. More privileged men may experience anger at the loss of access to women's bodies as social movements, such as #MeToo, and legal advances have helped women reclaim some of their bodily integrity (Messner, 2016). Even as masculinities are melting, shifting, and are being challenged and redefined, the narrative of "man heroically transcending and triumphing over a nature" involves control, violence, risk, and struggle (Beauvoir, 1968) and still permeates Western culture. It is this fusion of violence against nature and hegemonic masculinity that I turn to next.

Hegemonic masculinity as anti-ecological

While masculinities vary tremendously across time and place, masculinist traits are typically expressed as being in opposition to female traits (Plumwood, 1993). Further, masculinity is often regarded as superior to both women and nature (Kheel, 2008). It is these two features of masculinity: opposition to feminine traits and superiority over nature that binds and shapes a worldview toward both the more-than-human-life world and human–animals. Marti Kheel (2008) exposes the anti-ecological underpinnings of hegemonic masculinity. In particular, Kheel (2008) isolates the masculinist theme of "transcending the female-imagined biological world" among holist philosophers (p. 3). In *Nature Ethics* (2008), Marti Kheel argues that hegemonic masculinity is built on men transcending nature, the body, and those persons, both human and nonhuman, deemed to be lesser through their associations with the natural and the animal. Kheel (2008) argues that masculine identity may be a major contributor toward violence against nature. The inability to live up to masculine ideals may lead to violence when men feel unable to stake a claim to a masculine identity through "legitimate" means (Hunnicutt, 2009).

The cultural motif of the heterosexual white male ritually performing masculinity and achieving "manhood" by dangerous contests with nature is replete in Western society (Evans, 2002). Images of political candidates who are running for office commonly portray themselves as hunters, or stage hunting events during the course of a political campaign. The symbolic message of a man with a gun hunting in the woods communicates hegemonic masculinity, self-sufficiency, and the ability to tame and conquer "wild" elements. During the 2004 U.S. Presidential election, John Kerry, a Democratic Senator from Massachusetts, who unsuccessfully made a bid for the presidency, staged an elaborate, highly choreographed Geese hunting trip. Even as Kerry was not a hunter, posing as one was a strategic deployment to (re)claim entitlement to nature as the province of privileged, white, heterosexual males.

Consider how often we see the "man triumphs over nature" motif in contexts where men aspire to power and dominance, where the symbolic displays communicate that nature must be conquered, often violently. In Homer's The Odyssey, Odysseus, facing his greatest peril, says, "I will stay with it and endure through suffering hardship, and once the heaving sea has shaken my raft to pieces, then will I swim". How often in the archetypal hero's journey is a man violently struggling against nature—and how does that archetypal struggle come to define our gendered existence? In a poem by D.H. Lawrence, The Snake, the narrator admires and fears the creature and out of fear tries to kill it. The theme of men subjugating a wild, uncertain, threatening, formidable nature is a central theme in our Euro-Western cultural imaginings of masculine dominance.

The origin of dominating animals is an ancient mechanism of transcendence for males. Joseph Campbell (1991) notes that in Celtic myths men follow animals into the wilderness to find lands which lead to transformation. But men in these same ancient myths practice animal slaying as transcendence—a way to transform consciousness. Even the very act of meat-eating is a performance of masculine self-affirmation (Adams, 1990). In these instances, we see males seeking an ascent to a higher iteration of manhood or a declaration of gender by subduing nonhuman animals.

Lynn White (1995) and Rosemary Radford Ruether (1983) traced the roots of our culture's tendency to exalt masculine attributes over the characteristics of women, children, the earth, and nonhuman animals. Christianity's emphasis on anthropocentrism rather than locating the sacred in the earth, dominion over nature, placing hell beneath the earth, and worshiping a sky god, have all contributed to a patriarchal social order and anti-ecological masculine configurations. Prior to monotheistic, patriarchal religions that worshiped war-like sky god archetypes, value was instead given to nature and its cycles, fertility, nature, and women (Gaard, 2017).

The category of "animal" is foundational to understanding hegemonic masculinity. Throughout history, particular human populations have been described as more bestial or less "evolved" than others. Classifications of inferior or "animal-like" are bound up with notions of savagery, cognitive and physical ability, dependency, and even sexuality. The dividing line between animal and human has been embedded in gendered and racialized debates where assumptions about human traits have been used to classify animals (Taylor, 2017). Animal–human relations might be evoked in the portrayal of cultural difference as well. Politically motivated productions of hierarchies based on so-called "civilized" societies may be equated with greater distance from the "animal" (Seager, 2003). This scheme of differentiation is relevant to hegemonic masculinity because part of the narrative structure of masculinity—as a diffuse worldview—involved transcending the natural world.

Existing hierarchies that place humans above animals and men above women have greatly influenced how we classify species (Cochrane, 2010). Our classification of animals is not entirely informed by biology but is also socially determined (Bekoff & Pierce, 2017). The categorization of nonhuman animals as inferior goes hand in hand with the practice of dehumanization. And the practice of dehumanization exposes a plethora of bigotry: racism, sexism, ableism, heterosexism, and so forth. In this anthropocentric system, the world is the domain of "man" while animals (and human populations closely associated with the animals) inhabit a lesser, subordinated realm.

Hegemonic masculinities are anti-ecological (anti-life), and it follows that masculinity-as-dominance may be reaffirmed through violence. Marti Kheel (1995) examines the rather romanticized writings of hunters and detects not only sexual overtones but also the sentiment that the activity of hunting is important in the development of "manhood". It is possible that wild animals represent longings, urges, and needs that align with particular masculinities throughout history. In the hunt for masculine self-identity, violence is a key part of hunter's ethical code, albeit a restrained, denied, or renamed violence (Kheel, 1995). Kheel (1995) goes on to point out that this masculine identity strives to seek expression in *opposition* to the natural world.

Meat and its symbolic link to masculinity

Throughout human history, meat-eating can be traced along lines of power, reinforcing class distinctions. The European aristocracy consumed copious amounts of meat while commoners ate primarily second-class foods: grains, vegetables, and fruits, which are also considered "women's food" in many parts of the world (Adams, 2011, p. 36). While meat-eating is primarily bound up with affluence, Carol Adams (2011) details how meat-eating also establishes patriarchal distinctions. Adams (2011) uncovers the masculinist privileging of meat-eating by identifying numerous examples of how meat consumption is bound up with male identification, status, privilege, and importance. Men are more likely to receive meat in societies where poverty results in a deliberate distribution of animal flesh (Adams, 2011). In affluent societies, where meat is abundant, the diets of men and women become much more similar (Adams, 2011).

It is important to keep in mind that while the term "meat" strips the word of its violent context, that reference to meat consumption is fundamentally about killing. It is not possible to have meat (yet) that is not dead, that was not preceded by killing of and suffering by a sentient animal. Keeping this "context of killing" in mind, to the extent that meat-eating is a symbol of virility and male power, the consumption of meat also falls along racial and class lines. As Adams (2011) describes it: "The hierarchy of meat protein reinforces a hierarchy of race, class and sex". When the myth that meat is proclaimed to be the superior protein source, indigenous cultures whose dietary history favored vegetable and grain protein sources are subordinated. As societies maintain power arrangements and customs that favor white people, the consumption patterns of white people—which include habits of meat-eating—are imposed on people of color (Adams, 2011). But even then, people of color are given the lowest quality meat available. For example, in contemporary times, African Americans have been targeted as key consumers in the explosion of fast food chains in the USA. Indeed, the U.S. Government has subsidized fast-food outlets in minority communities (Jou, 2017). The result is that minority communities have suffered some of the worst health consequence of consuming cheap, low-quality, convenient fast food which has contributed to still growing rates of obesity, heart disease, and diabetes (Jou, 2017). Richard Twine (2014) continues this line of analysis by exploring how animality has been used to mark both race and nation through food colonization and global neoliberal capitalism.

We can see the association of masculinity with meat by the linguistic associations we make between the two. One of the definitions of meat, according to Merriam-Webster, is the core of something. This meaning is evident in phrases such as "beef up" or "the meaty part of the story". Meanwhile, among the many definitions of vegetable includes a person having a dull or merely physical existence, a person whose mental and physical functioning is severely impaired, and someone who requires supportive measures (such as mechanical ventilation) to survive. Thus, vegetables are considered passive and meat is active and considered a central part of something. It is no coincidence then, that as vegetables are considered "women's food" they by extension become passive and lifeless and men's need to distance themselves from them exposes the sexist ideology embedded in the gendered patterns of foodways (Adams, 2011).

Meat is a symbol of male dominance, which is evident in the belief that eating the muscles of animals will make us strong. Adams (2011) identifies this superstition as part of a patriarchal mythology where the foods associated with second-class citizens (i.e. women the gatherers) are considered second-class sources of proteins, despite evidence to the contrary (Harper, 2010). Beef, in particular, has a strong association with masculinity, which is evident in everything from advertising to colloquialisms, sports, and art (Rogers, 2008). That meat-eating is a masculine activity is evident in cookbooks that address the barbeque sections to men (Adams, 2011) but is also evident in meat advertisements that link meat consumption to a hypermasculine subject. Rogers (2008) points to the dualistic oppositions in play with beef advertisements where men are pitted against women and meat versus vegetables. Rogers (2008) uncovers meat's gendered symbolism to show how meat consumption is a way to validate masculinity, and an opportunity to provide men with symbolic compensation when other attempts at performing masculine scripts have failed. In this sense, meat-eating provides a sort of reassurance of their maleness.

While the history and patriarchal implications of "the hunt" were detailed in the previous chapter, it is worth mentioning here that Adams (2011) also finds agreement with these scholars as she traces the association with meat as a symbol of male dominance. Gender inequality, according to Adams (2011), is bound up with species inequality because in many cultures meat was procured by men through hunting. A valuable commodity, the men who controlled this meat procurement held power as a result. Adams' (2011) argument is supported by anthropological research by Peggy Sanday (1981a) who, in a survey of over 100 non-technical societies, found that plant-based economies were more likely to be egalitarian where women enjoyed more autonomy and respect for their work in providing plant-based protein sources. Conversely, Sanday (1981a, 1981b) found that meat-based economies were more patrilineal, segregated by sex, with women doing more work than men, were responsible for childcare, and women's work was less valued.

Evidence abounds that meat is a symbol of male dominance. "The beast" is a common trope that signifies both positive and negative forms of "primitive" masculinity (Rogers, 2008). Men who choose vegetarianism are emasculated. While Rogers (2008) recognizes that there are many different metonyms for hegemonic masculinity, eating meat, as it is presented in commercials advertising fast food which references a primitive, bestial masculinity, offers the possibility of one's masculine status being symbolically restored. Meanwhile, vegetarianism is coded as the as both feminizing and emasculating. Therefore, vegetarianism is a threat to hegemonic masculinity.

Masculinity and blood sport

Blood sports are a mechanism of entertainment that involves the violent performance of "human against animal" often as competitive spectacle. That animals are the targets in blood sports is contingent on a speciesist framework. Speciesism is rooted in the legal traditions of liberal democracy where human rights were predicated on human exceptionalism (men) (Ryder, 1989). This "human exceptionalism" was derived from either rationality, the soul, or the "dignity of man" and originated from Christian notions of the superiority of humans. The exalted status of the human influenced the secular sphere and spread across the world to shape our current global political configuration. One of the ritual enactments of this gendered speciesism is through the violent subjugation of animals through sport (Lawrence, 1982). This form of animal oppression is linked to patriarchy and dominant constructions of masculinity which find affirmation in subjugation of nature.

Serpell (2001) indicates that our culture has a 700-year history of practicing animal abuse in still thriving traditions such as rattlesnake roundup, rodeo, and sport hunting. These practices are widely condoned. While factory farming is not a sport, the widespread acceptance of this institutional practice of captivity, breeding, abuse, suffering, and neglect occupies the same ethical realm about concerns over how animals are treated in both zones of human consumption and entertainment. Blood sports such as cockfighting and dog fighting are less normalized, largely because these practices are linked to the devaluation of marginalized social groups (Cook, 1994).

Blood sports, bullfighting, and cockfighting, for example, are fundamentally rituals about masculinity, sex, and the display of animal aggression (Kalof, 2007) where vanquishing a "worthy foe" and a "worthy kill" demonstrates mastery (Emel, 1995). Such male-focused sporting rituals reflect the masculine values of sexual potency and aggressiveness. Take the phrases, "riding the bull" or "grab the bull by its horns", which conjure up an image of a (male) matador achieving superiority over a strong, formidable, but inferior nonhuman animal. Perhaps fears of all things "wild" and "nonrational" manifest in defeating and breaking animals.

Meanwhile, ferocious animals may be both greatly valued and feared. To the extent that some animals embody courage and valor, they may be acquired, killed for the trophy, displayed or tamed as a symbol of patriarchal dominance. Roman emperor Augustus used exotic animals in gladiator rings to distract the Plebeians from the expanding empire and also sacrificed animals in public rituals. Warriors, generals, and soldiers of the Roman Empire may have had fierce, but valorized animals, such as lions and jaguars, as pets. The kings of Mani Kongo, now present-day Congo and Angola, which existed from the 14th to the 19th centuries, carried a whip made of zebra tail and from their belts hung the heads of baby animals—both symbols of their authority (Hochschild, 1999).

The gendered performance enacted through sports serves as a validation of masculinity and sexual virility. Consider the incident of an Iowa coach who spray-painted a live chicken gold. The gold represented the golden eagles of a rival team. The coach then threw the chicken—who was alive and terrified—onto the field and instructed the players to kick the hen around. The name of the game was "Get the Eagle" (Ruth, Gilbert, & Eby, 2004). This particular instance shows how inseparable masculine dominance is from the mistreatment of Others—in this case, a nonhuman Other for whom there are almost no regulations for their humane treatment (Francione, 2008).

In writing about cockfighting, a decidedly male event, Clifford Geertz (1973) argued that roosters who fight are actually extensions of the male ego. Researchers have discovered connections between cockfighting and subsequent worldview that embraces dominance, aggression, violence, authoritarianism, cultural patterns of hostility, and homosocial male bonding (Cook, 1994; Dundes, 1994; Hawley, 1993). In describing the hyper-masculine culture of blood sports, Kalof and Taylor (2007) write that blood sports are:

> ...symbols of a culture infused with macho aggression and menacing violence. Both cock fighting and dog fighting are sport activities staged by humans in which animals are incited to fight, maim, and kill each other. Both are focused on competition without a survival-of-the-fittest component; winning as a singular goal with little interest in the process of fighting, only the outcome; spectators who watch the fights and validate the superiority of the winning animal's human handler; and

gambling on one of the animals to win (Cashmore, 2000). Further, in both sports there is a clear juxtaposition between owning fighting animals and aggressive masculinity.

(p. 321)

Jody Emel (1995) analyzed the near extinction of wolves in the USA as a consequence of the normalization of hunting and ultimately a distortion of male power. Wolves were killed for land, investment, pelts, to keep big game animals alive so that human hunters could continue their "sport", data and for trophies. But long after wolves were a threat to economic investment, they continued to be poisoned, set on fire, tortured, shot, and annihilated. Emel (1995) showed that the mass killing of wolves was due to an

intertwining causality stemming from a dominant construction of masculinity that is predicated upon mastery and control through the hunt. A fear of the "wild", the "irrational", or the "different" is also part of the construction. Wild animals, and particularly predators such as the wolf, have represented longings, needs, and urges that were suppressed in the particular construction of masculinity that dominated during the late 19th century and early 20th century. They have been targets for hatred, the same hatred that launched armies and lynch mobs against human "others".

(p. 720)

The extermination of these "fierce animals" was an almost superstitious performance of virility and manhood within a mythical context of the American frontier imagination. Emel (1995) goes on to situate this brutality as a construction of masculinity which is fed by class dynamics, cruelty, bureaucracy and rationalization, and commodity production.

The racialization of masculine identities through violence against animals

Ecofeminists have pointed to the intersections of animal exploitation and racism (Adams, 1994; Deckha, 2012; Twine, 2010) and while there are a number of ways that masculine identities are racialized through the violent subjugation of nonhuman animals (e.g. the slaughterhouse), blood sports are particularly charged issues. Blood sports such as dogfighting and cockfighting are very much bound up with race and class. These blood sports are no longer practiced by the upper classes but are now found in communities of poor whites, blacks, and Latinos. Because these blood sports are now illegal and are associated with socially marginal groups, the discourse surrounding dogfighting and cockfighting are regarded as the domain of "gangsters and drug dealers".

While dogfighting and cockfighting are sites of racialized masculinities, there are many instances where race is stripped from analysis, as in the highly publicized dog fighting case of Michael Vick. Michael Vick, an African American star quarterback for the Atlanta Falcons, was eviscerated by the public and press for running a dogfighting ring. But the discourse surrounding Vick's participation in dog fighting often proceeded in a context as if race did not matter. Patricia Hill Collins (2006) refers to this phenomenon as "new racism", which is a period marked by discourses of race neutrality (color-blind transcendence) in contradiction with continued practices of racialization. In order words, race is continually denied even when our world is saturated in it. As much as we may deny its presence, there is no such thing as a context-free from race. Vick was indeed extremely cruel to his dogs, but I point to the larger context of race here to show how animal cruelty is much more socially acceptable when it is practiced by middle and upper classes. As a black man, even as he was wealthy and successful, the public vilified Vick for his participation in dogfighting. Meanwhile, other forms of animal cruelty, which may involve just as much anguish for the animals, are left unchallenged (i.e. horse racing, horse-drawn carriages for tourists, circuses, vivisection, factory farming). Or as Breeze Harper (2011) puts it,

> How is covert "whiteness" in the United States maintained by sensationalizing and reprimanding DMX and Michael Vick for animal torture and cruelty, while ignoring the "animal gaming" pastimes of white privileged males, casting them as nondeviant "normative" behaviors?... white male youths who kill "game" animals are "heroes"; black male youths who kill a puppy or engage in dogfighting are "evil." As a scholar, I wonder, are speciesism and racism in the United States contingent on each other?
>
> (p. 74)

Exposing animal cruelty in non-white populations is a delicate matter as we run the risk of using speciesism to reproduce stereotypes of minority groups as "backward". Examples abound of animal rights activists accusing marginal groups of cruel practices toward nonhuman animals and in the process positioned such groups as a threat to White identity. Since whiteness and speciesism are the norms, such normative beliefs shape our beliefs, relationships, and reactions with human and nonhuman animals.

Kalof and Taylor (2007) report that dogs who are manipulated to fight occupy important signifiers in gang culture. For gang members, their dogs' are emblems of status. A dog might be matched to a particular member as an indicator of authority and influence in the gang but might also be matched with the intention of intimidating other gang members. Kalof and Taylor (2007) point out that dogfighting is a lucrative organized crime activity, but it is also connected to a host of other crimes involving theft, drugs, and gambling. Moreover, if a dog fails to perform in the pit, they may be killed by their "masters" out of embarrassment for their cowardice (Gibson, 2005). Existing scholarship on dogfighting seems to point to a broader culture of violence, one that is characterized by the violent control of the other-than-human-life world. In contrast to Kalof and Taylor (2007), Evans, Gauthier, and Forsyth (1998) reported in an ethnographic study that when White, working-class men participated in dog fighting, the practice provided affirmation of their masculinity where fighting dogs were imbued with characteristics of mythic heroes, such as bravery and skillfully defeating an opponent. This enactment of mythic masculinity through the fighting rituals is also similar to expectations for human males who are also expected to fight bravely and may face a punitive end for acting cowardly.

In the process of colonialization, the racially unmarked category of "white" is equated with "civilized" and "human" while drawing on a speciesist ideology of animal inferiority (Twine, 2014). Meanwhile, a host of historical violences—genocide, slavery, colonization—have gained currency by racializing the "other" as animal. The link between exploited animals and racism is a complex terrain. Kim (2007) examines immigrant animal practices that have become tense political struggles, focusing in on California, in particular, due to its large Mexican and Asian immigrant populations. What many would view as a violent and cruel practice, horse tripping, was banned in 1994. Some Mexican immigrants and their descendants have carried on this practice, which has been carried over from the ranching practices in the Mexican hacienda system. These contests involve a cowboy (Caballo) riding one horse, chasing another horse while seeking to lasso the horse's front legs which "trips" the horse again and again until the animal is defeated with injuries or death. If the horse survives, she is shipped to slaughter after the contest concludes (Barraclough, 2014). This Mexican rodeo tradition has produced conflicts over animal practices that have been used to further marginalize Mexican immigrants. Barraclough (2014) exposed the subtle dynamics of power involved in the criminalization of Mexican rodeo and the ways this struggle contributed to the marking of immigrants as "illegal".

The laws against horse tripping and the media coverage of the struggle served to bolster demands for a militarized border, and greater exclusionary and deportation practices of Mexican immigrants. The sometimes indirect and inadvertent fortification of anti-immigrant sentiment has led some advocates of immigrants to evoke a multiculturalist interpretive framework in defense of these animal practices (Kim, 2007). Of course, it is often the case that violent practices toward both human and nonhuman animals were acquired through imperial and colonial practices (Smith & LaDuke, 2015). But that longer history is often lost when interventions on behalf of animals merely advance long-standing practices of White aggression against people of color. As Kim (2007) writes:

In this view, the majority is guilty of judging immigrant minority cultures to be deficient and wrong (ethnocentrism), seeking to impose dominant cultural values on marginalized cultures (cultural imperialism), and enunciating an impassable racial difference between the "self" and the "other" in order to bolster ongoing efforts to exclude the latter from meaningful membership in this nation (racism and nativism).

(p. 234)

In these political tangles, it appears that animal and anti-violence advocates are further aggravating oppressive power over marginal groups, possibly perpetuating the very aggression we seek to remedy. The killing of animals becomes racialized when a value distinction is made between "sadistic" individuals who kill animals in ritualized performances, versus "civilized" folk, who still kill animals, but do so through an elaborate industrialized apparatus. This is where both an "ecofeminist reflexivity" (Twine, 2014) and a careful balance between protecting immigrants and people of color from cultural imperialism come into play (Kim, 2007). But at the same time, it is also crucial to maintain moral standards against violence and cruelty that reach beyond anthropocentrism but also accounts for the complexity of power relations and cultural contexts (Gaard, 2001).

Rodeo

Rodeo is a blood sport that dates back to cattle herding in various countries that positions the cowboy (mostly men) in competition with horses and cattle. Roping, wrestling, tying, racing, and riding a bucking bronco are some of the competitive activities in rodeo. Butterwick et al. (2011) found that rodeo is one of the most dangerous sports in the world, with catastrophic injuries exceeding that of football. Results showed that roughly 20 of every 100,000 rodeo contestants will experience a catastrophic injury. In this study, catastrophic injuries were defined as a rider who either died or had their life altered in a profound way. By comparison, in football, that rate of catastrophic injury is less than one in every 100,000 players. Rodeo is also egregiously cruel to the animals and has caught the attention of animal rights advocates. For example, one of the day-to-day violent activities in rodeo is branding—the burning of cattle and horsehide with a red-hot iron. Animals are chased, roped, tied, and burned. Empathy for the bawling, bellowing, squirming animal would interfere with the deed (Lawrence, 1982). Indeed, rodeo is banned in select locales in the USA and is completely banned in both the Netherlands and the UK (Morgan et al., 2001).

Rodeo is a blood sport that carries on the practice of violence within a frontier mystique. Lawrence (1982) suggests that the killing of animals in rodeo is a practice of almost extracting power from nature. For example, horses possess grace, strength, and speed—qualities which men lack, but admire.

The "vanquishing" of these animals is a violent conquest in which the conqueror believes he is acquiring the power from the animal by "taming their spirit" (Lawrence, 1982). In this sense, animals in blood sport become sacrificial animals:

> much of the cowboy's equipment, derived from the Spanish, is punishing to the horse. His stock saddle is heavy beyond utilitarian purpose, his spurs are designed with elaborately pronged rowels, and the bridle bit is of a type that has been described as an instrument of latent torture.
>
> (p. 157)

Lawrence regards rodeo as a ritual performance that addresses the dilemma of man's relationship with nature. Rodeo aims to manipulate and transform nature, while dramatizing and glorifying the practice of "taming the wild". The masculine performance of rodeo is dripping with dualisms of wild/ tame, male/female, predator/prey, nature/culture, and nondomesticated/ domesticated.

Elizabeth Lawrence (1982), in an extensive ethnographic study of rodeo, found that women and livestock are interchangeable in rodeo culture. Rodeo slogans on stickers and signs reveal the way that women and animals become interchangeable, denigrate, and eroticized: "Steer wrestlers get it on the side", "Ropers handle anything horny", and "Bull riders ride the wild humpers". A strictly masculine practice, women's participation in rodeo is very limited, mostly to barrel racing, which does not include human–animal antagonisms. When interviewing rodeo contestants, Lawrence (1982) uncovered widespread hostility toward women's liberation and strong opposition to women participating in the sport, citing weak bodies:

> Women as brainless sex-object and made closer to elemental nature by being characterized as a child, is a frequently repeated theme in rodeo. It leads quite logically to the notion that women, since they are juvenile, should and indeed must, be dominated by men, along with the rest of nature. The process of conquering a woman has its metaphoric counterpart in the male conquering of the West itself. The cowboy then becomes the archetypal representation of the male element in the conquest of virgin land.
>
> (p. 114)

Lawrence (1982) deduced that the display of various cowboy symbols was related to his role in conquest—of the land, the wild, animals, nature, his own fears, and women. American novelist Larry McMurtry, who spent a good deal of his life in the company of Texas cowhands, said that their working life was full of violent activity and expected that such violence would spread from animals to humans. He said that he has never known a cowboy to be a gentleman.

The fusion of objectified animals and sexualized females

The human-centered hierarchy is revealed in our preference for certain types of nonhuman animals and our vilification of others. We are much kinder to animals that resemble a human ideal of beauty—animals that closely resemble human babies, such as baby seals and panda bears—wearing an expression of doughy features, liquid eyes and lugubrious looks. A mark of patriarchal society, looks matter. For females and animals, the relative beauty of animals/nature serves as a threshold for destruction. Colorful fish are pleasing and end up staring in animated movies. As attractiveness is cultural currency for both women and animals in a patriarchal culture, it follows that threatening animals and insects with alien looks (flies, scorpions, reptiles, invertebrates) are either tokens of exotic display and possession or are destroyed. Warthogs and rhinos with disproportionate features and callous, warty skin, rather than smooth baby-like skin, or soft, furry coats occupy a lowly status in the nonhuman animal hierarchy.

Inspiration for monsters, vampires, and aliens come from repulsive animals or creatures that only come out at night, revealing our fears of animals as "killers". Fangs are sinister and threatening, symbols of poisonous species who we zealously eradicate. Wolves are portrayed as conniving killers in children's fairy tales, such as the Three Little Pigs and Little Red Riding Hood. Parasites and worms are symbols of invasive agents that must be exterminated and managed. Rather than allowing each life form in the biosphere to fulfill its occupation, patriarchal culture seeks to assign value, eliminate, manage, sort, and control. There is hardly an area of human culture that is uninfluenced by our relationship with the more-than-human-life world. The activity of human–animals who name, classify, decide, dissect, evaluate, monitor, cage, and observe, reinforces this hierarchical relationship of humans above, nature below. One of the hallmark manipulations of nonhuman species is our culture's reliable fusion of women and animals.

Adams (2011) traces the fusion of women and animal subjugation to the fall of humanity as outlined in the Judeo-Christian creation myth where Eve and the serpent were blamed for the fall (Seibert, 2012). Adams (2011) argues that there is a link between violence against women and animals through the fused categories of gender and species. The human-centered hierarchy requires a distancing and denigrating of animal others. Further, we equate animals to something already objectified in order to emphasize the difference in social value. For example, referring to particular social groups as "animal-like" achieves three outcomes: the reproduction of human domination, reinforcing the disposability of animal bodies, and the legitimation of oppression toward those humans considered "animal-like". Adams (2011) wonders if metaphor itself could be the "undergarment to the garb of oppression?" (p. 58).

The conflation of women with animals might be in entertainment displays and is often deployed in erotic ways. That discourse surrounding sex is often laced with violent imagery is commonplace (Kalof, Fitzgerald, & Baralt, 2004). Consider slang words for intercourse. Many such words are forceful words, such as bang, pump, and screw. Just as we might fuse violence with the erotic, we "animalize" women while simultaneously eroticizing them. The iconic playboy bunny is a major cultural reference point for this sort of practice. It is not surprising then, that we might see scenarios where eroticized violence includes animals. Examples of nonhuman animals being violently controlled in highly masculine ways abound in our culture and are sometimes fused with the erotic. On April 20, 2010, the U.S. Supreme Court struck down a law banning animal cruelty videos, granting free speech to people who make them. The ban was reinstated when President Obama signed the Animal Crush Video Prohibition Act into law in December of that same year. "Crush" videos are recordings where women in stiletto heels or bare feet stomp, crush or impale small animals. The eroticized display is intended to satisfy the sexual fetishes of sadistic viewers (Animal Welfare Institute, 2018). In this example, we see implements of violence (the feminized shoe) seeking to control and destroy a helpless body and where the act of violence and women's body parts (feet) are fetishized.

Meat metaphors that are frequently evoked following violence (particularly against women) suggest that rape, in particular, is very much related to the violation, fragmentation, butchering, and consumption of both animals and women (Adams, 2011). Adams (2011) offers an extended analysis concerning the overlap of butchering animals and rape of women:

> Rape, too, is implemental of violence in which the penis is the implement of violation. You are held down by a male body as the fork holds a piece of meat so that the knife may cut into it. In addition, just as the slaughterhouse treats animals and its workers as inert, unthinking, unfeeling objects, so too in rape are women treated as inert objects, with no attention paid to their feelings or needs. Consequently, they feel like pieces of meat. Correspondingly, we learn of "rape racks" that enable the insemination of animals against their will. To feel like a piece of meat is to be treated like an inert object when one is (or was) in fact a living, feeling being.

In contrast, when the metaphor "rape of the earth" is used loosely, the "absent referent" then becomes violence against women. When the figurative use of the word "rape" is applied to other contexts, the actual lives of those who have been raped are elided. Berman (1994) offers caution about using the term "rape" to signal the exploitation of nature because in the process we make rape survivors invisible (absent referent) and simultaneously reinforce sexual violence. Berman writes:

The absent referent can be found in many metaphorical sayings which link animals and women—women have been referred to as cows, dogs, bitches, beavers, bunnies and finally "pieces of meat". What is absent in these sayings is the woman herself and the violence that underlies these derogatory terms. Through the absent referent the subject is objectified, and patriarchal values are institutionalized.

(p. 176)

When survivors of violence say that they felt like "a piece of meat", they are evoking a system of language in which metaphors reference the original fate of the animal who was violently deprived of life, without regard for animal's own desire to be free from suffering and harm. To feel like a "piece of meat" is to be regarded as less-than-human, which is "matter without spirit" (Griffin, 1979).

Carol Adams (1996) details cases of abuse and battering against both women and their pets. The joint abuse of women and animals reveals the fused oppression between women and nonhuman animals, but also how animals are used by a batterer to control women by creating a climate of terror or using threats to kill the family pet to ensure silence among the battered woman. In violent domestic situations, animals may be employed as tools of psychological and physical terror as objects against which humans vent rage. Adams (1996) employs the concept of somatophobia to explain woman–animal abuse connections. Somatophobia is hostility directed at the body, according to Elizabeth Spellman (1982), and is an outgrowth of sexism, classism, racism, and speciesism. This hostility is reserved for particular kinds of bodies, those abject, subjugated, "othered" bodies, namely, animals, women, nonnormative genders, people of color. In a bizarre nexus of somatophobia, speciesism, and homophobia, the Bolivian President, Evo Morales, in 2010 attributed the cause of homosexuality (distorted bodies) to the consumption of genetically modified chicken (poisoned, altered, violated bodies): "The chickens, they deviate from themselves as men" (ILGA, 2010).

Adams (1996) provides empirical evidence in the form of testimony from survivors and their advocates which reveal key patterns that expose connections between the abuse of women and the abuse of animals. These connections point to a kind of patriarchal militarism surrounding relations between humans and animals. There is the killing of an animal, usually, a pet, which Adams (1996) claims is performed to establish control over women and children who are already being abused. Next, Adams (1996) explores cases where animals are used in sex acts along with violating women or children. Finally, Adams (1996) reports anecdotal evidence which indicates that child victims of sexual abuse may go on to injure animals, revealing that sexual exploitation influences behavior toward animals.

Taking seriously violence against nonhuman animals

These various ways in which animals are subjugated are bound up with forms of patriarchal domination. The gendered foundations of industrial and "recreational" animal abuses is shocking in its horror. Yet it is taken for granted in a cultural context where animal abuse is casual and routine. To ignore the role of animal others in social life is also to fail to understand our constructions of self, our moral foundations, and varied manifestations of violence (Emel, 1995; Jenkins, 2012). Taking seriously the emancipation of nonhuman animals is one part of a larger effort to dismantle structures of oppression (Beirne, 1999; Gaard, 2012). The dominances of nonhuman species are embedded in our social order with legal, social, and economic logics in place that perpetuate speciesism and institutional forms of violence toward animals (Gruen & Weil, 2012b). This complex apparatus of "biopolitics" involves the subtle practice of regulating bodies through numerous techniques that subjugate and control (Shukin, 2009; Siisiäinen, 2019). Aristotle argued that language separated humans from animals, which perpetuated the belief that nonhuman animals did not possess language. Of course, animals do communicate, but our own speciesism assumes that human language is superior. The belief that our capacity for spoken language permits us to look down upon animals is the same ideology that legitimated racism toward humans who spoke "primitive languages", particularly in the 19th century. In such instances, racialized Others were therefore regarded as "savage people" and "inferior" and "evolutionary throwbacks" resulting in violence stemming from such dehumanization (Van Cleve, 1993).

The domination of animals almost always points to the "superiority" of human traits—complex emotions, the capacity for rational thought, bipedalism, opposable thumbs, and language. Rendered inferior, animals can then be exploited in a sort of "morality-free" zone and those humans who have been closely associated with animals (women, people of color, cultures with "primitive language" homosexuals, intellectually and physically disabled) may also be dehumanized, also making exploitation more permissible.

Race, like gender and other markers of difference, are entangled with animality. Animals, women, and people of color have been compared in disparaging ways throughout history in a variety of contexts, pointing out that there is indeed a shared ideology that oppresses both humans and nonhuman animals. This is not to say that animals and marginalized humans experience oppression in the same ways. In examining the relationship between the violence and subjugation between human and nonhuman animal bodies, we can't necessarily assume shared experiences of exploitation. Instead, we must examine the specific contexts in which such exploitations play out. Indeed, these experiences of domination play out in wildly different ways. But speciesism feeds the same ideologies that regard some human lives as less worthy. The "inferiority" of women and people of color is commonly dehumanized and animalized, pointing to the "animal nature" of such groups, pathologizing entire populations as having diminished worth.

Adams (2011) unearths numerous instances of feminist scholars pointing to the intersection of the oppression of animals and the oppression of women. Patricia Hill Collins (2006) in *Black sexual politics: African Americans, gender, and the new racism*, writes on the topic of treating women like animals: "The truth is in sexist America, where women are objectified extensions of male ego, black women have been labeled hamburger and white women prime rib" (p. 112). But in statements such as these, Adams (2011) points out that this literature seeks only to advance women's issues, leaving the fates of animals forgotten. In this way, these very feminists are colluding with a patriarchal culture. Understandably, we might use butchering of animals as metaphors for violence against human animals. But we must proceed in such a way that also acknowledges the otherwise "absent referent" of the nonhuman animal—to make visible the violation of animals as well.

The core theoretical concept in Adams' work is the absent referent (2011). Adams details the process by which living animals become absent referents through their conversion into "food". Their bodies are transformed through butchering, and the dead animal is then renamed as "meat" and simultaneously made absent. When humans use animal metaphors to illustrate their own human experience, animals are also made into absent referents. A common animal metaphor is used with regard to meat metaphors used to describe women. Adams (2011) continues:

> When radical feminists talk as if cultural exchanges with animals are literally true in relationship to women, they invoke and borrow what is actually done to animals... They butcher the animal/woman cultural exchanges represented in the operation of the absent referent and then address themselves solely to women, thus capitulating to the absent referent, part of the same construct they wish to change. What is absent from much feminist theory that relies on metaphors of animals' oppression for illuminating women's experience is the reality behind the metaphor. Feminist theorists use of language should describe and challenge oppression by recognizing the extent to which these oppressions are culturally analogous and independent.
>
> (p. 72)

Adams (2011) goes on to hold animal advocates accountable for employing language that evokes rape as a metaphor for violence against animals. It is important to acknowledge the reality behind the metaphor. In other words, if we call forth metaphoric language, we need to also expose the original harm. The goal of anti-violence scholars and activists should be to eliminate the very structures that create absent referents, while simultaneously acknowledging intersecting oppressions.

The twin crises of human and nonhuman animal oppressions mean that there is no gender justice without interspecies justice. There is a joint need to simultaneously challenge the devaluation of animals and the earth and dehumanization of the Other. For some time, feminisms' focus on economic injustice elided other injustices but, over time, once contentious and less visible issues have, imperfectly and disparately, become a part of mainstream feminisms (Fraser, 2009). While the scope of acknowledged injustices has widened, there is still very much a selective incorporation with regard to species in academic feminisms. Mainstream feminisms' visions of a just society are anthropocentric in the sense that nonhuman animals are infrequently included and nonhuman animal violences are rarely given attention beyond their implications for human–animal persons. If the aim of feminism is to fully *dismantle* systems of human oppression and violence, then this promise must be extended to nonhuman animals as well. While feminist theorizing has the potential to create remarkable, radical social, political, and economic change, efforts to undo the status hierarchy has historically left out one crucial piece: species.

While still a minority of voices, there are several scholars who have stressed the importance of taking up the animal question (Birke, 2002; Deckha, 2006, 2013; Donovan, 1990; Gruen, 1993). Despite a substantial minority of feminist scholars who have argued that feminism should take up critical animal studies, the topic rarely makes it into major feminist journals, texts, and conferences. There are several reasons why feminist scholars have avoided this issue. The call to adopt a vegan diet evokes a second wave solution of the "personal is political" (Hochschartner, 2014). The animal rights movement has been accused of not taking seriously issues of class, race, gender, and sexuality. Indeed, while ignoring issues of intersectionality, animal advocacy has often advanced a white, middle-class model. Some vegan movements have engaged in body shaming to get people to adopt a plant-based diet while advancing an image of a white, heteronormative, ableist image of how vegans should look (Harper, 2011; Taylor, 2017). Taylor points out that "most diet and fitness books presume a while straight person without disability—but there is something especially offensive about these tactics when they come from a movement that claims to value compassion" (p. 61). All that said, feminist interventions offer promise to destabilize harmful, patriarchal, Western consumption practices.

Of course, the actions of industry have a greater impact than individual consumers. While it is important to point out the ways in which consumption, ecological degradation and the effects of climate change are staggered by social class, it is equally important not to uphold a neoliberal agenda. It is common to encounter a focus on individual actions to mitigate climate change. This position implies that individual consumers are singularly to blame for what really is the irresponsible activity of industry and government regulations. Attitudes and behaviors are important. The problem of overconsumption is important, and individuals must care about our ecosystem for change to happen, but political and structural change is necessary to affect environmental degradation.

There is a fear that discussions around an "ethic of care" will return us to essentialist territory (Donovan, 1990). There is a fear of being charged with biological determinism (Twine, 2010), and feminists scholars might want to avoid any association with "nature" after so many decades of work to dismantle that link (Birke, 2002). If most feminist scholarship avoids animal studies, it is also true that most work on animal ethics fails to consider animals as situated in a larger economic apparatus (Gruen & Weil, 2012a).

Since this same logic of domination is expressed in sexism and the subjugation of nonhuman animals, it makes sense that feminist scholars would take up the animal question. Both the animal liberation movement and the anti-gender violence movement should centralize these oppressions as intricately entangled. Inattention to the intersection of gender violence and animality will leave anthropocentrism unchallenged. To the extent that anthropocentrism is available for use, it can be co-opted by other systems of oppression.

In writing on why feminism is perhaps best positioned to take on questions of the animal, Emily Clark (2012) writes:

> This is manifest in feminist theory's commitment to the materiality of the body, to attending to those bodies most vulnerable to abuse, to exposing the logic of exclusion and the politics of abjection, and perhaps most important, in the ethics of representation and "speaking for" others. In both theory and practice, feminism has the greatest capacity to take on what is one of the most ethically and intellectually challenging issues of our time.
>
> (p. 517)

Just as humans should be approached with an intersectional lens, we should think intersectionally about animals. The mainstream feminist academy and gender violence scholars should consistently deal with speciesism, as it is inextricable from human systems of subjection, exploitation, and violence.

Conclusion

It is not uncommon to see references to violence against animals in everyday life. Dog fighting videos are available on the internet. The control of nature and the challenge of taming its "wildness" with allusions to violent sport are evident in all sorts of symbols and discourse. The phrase "catch the big fish" is commonly employed with regard to drug business and criminal justice pursuit of drug lords—two hyper-masculine entities engaged in an often-violent enterprise of power and wealth. To "kill two birds with one stone" summons an act of violence against animals in a casual idiom employed to allude to efficiency. While such phrases masquerade as gender-neutral, they allude to a long history where, as Virginia Woolf (1938) described in Three Guineas, "Scarcely a human being in the course of history has fallen to a woman's rifle; the vast majority of birds and beasts have been killed by you, not by us" (pp. 13–14). In this chapter, I explored the link between masculinity and dominance, violence and aggression, hierarchy, and mastery over earth. While being careful to avoid simplistic dualisms, I considered how hegemonic masculine identity may be a major contributor toward gender violence.

Hegemonic forms of masculinity are rooted in ideologies of dominance over "nature", predicated on control and manipulation of animals through the hunt and blood sports. Blood sports are gendered rituals where toxic masculinities are enacted. These violent performances are predicated on the violent manipulation of nature (animals), control, dominance, and the obliteration of weakness (note that weakness is associated with femininity). This "logic of domination" over the environment extends to gender violence in multiple ways (Warren, 1996). The domination of emotions, which is necessary to carry out killing is evident in the violence toward animals detailed here. Hunting, in particular, is often "naturalized" by appealing to beliefs about the other-than-human-life world. The practice itself reveals a masculinist profile that seeks to demonstrate mastery over the self and the material world. Killing of nonhuman animals is a cruelty bound up with both powerlessness and power—a will to mastery. But cruelty toward nonhuman animals also invites other oppressions and brutality (Emel, 1995).

Chapter 4

Trans-species harm and the multiplicity of violence from climate change

Our relationship with the more-than-human-life world impacts gender violence. This relationship is multidirectional, particularly in the age of the Anthropocene where our drastic manipulation and destruction of the biosphere and its inhabitants have reached unprecedented extremes (Opperman & Lovino, 2017; Schwägerl, 2014; Seager, 1993). Just as our ideological, situational, structural, and relational stance toward animals and the earth is connected to human violence; the environmental crisis in turn affects our behaviors, adaptations, and responses, including violent ones. For example, global warming has resulted in an advance of the Saharan desert and subsequently violence in Darfur, which is patterned along gender lines. Having explored the patriarchal domination of "nature" in the previous chapters, I now consider how gendered violence is exacerbated by climate change and trans-species harm. Our masculinized "mastery" and control of the biosphere and nonhuman animals have culminated in our current climate crisis. As deforestation, drought, and disasters increase, climate-related conflict also rises (Adger et al., 2014). As violence toward nonhuman animals continues through the institutionalized terror in the factory farm industry, structural violence through climate change is intensified. Moreover, the trauma of animal execution likely fuels direct interpersonal violence, with gender-specific manifestations. Exploring the embeddedness of violence in climate change is like pulling on one thread and an entire cloth rises up. To organize these multiple threaded gendered violences, I address each in terms of direct, structural, and cultural violence (Galtung, 1969, 1990).

The scope of climate-related injustices and their racialized and gendered effects are vast, so I focus on just a select few here (Black, 2016). For example, it is important to explore how women suffer the worst effects of climate change and face unique risks of gender violence following climate-related disasters. I then move on to one of the major causes of climate change—greenhouse gas (GHG) emissions resulting from the exploitation of nonhuman animals for food in the form of livestock production and the extent to which nonhuman animals are oppressed by ideology rooted in the "exploitability" of bodies that are predetermined to have less value. I then move on to expose the direct violence against nonhuman animals in the factory farm industry. My aim here is to connect human and nonhuman oppressions, revealing their inextricable bind. I address the violence experienced by slaughterhouse workers, often the most marginal humans, recruited from "expendable" migrant labor pools. While targeting the impact of this cruel industry on human animals, I also consider the literature that explores the possibility that cruelty toward nonhuman animals might make violence between humans more likely. I am fundamentally concerned with the following questions: How might one type of violence morph into another? In what ways might violence against nonhuman animals bolster human-to-human violence? But these questions should be situated within a broader context of domination of the environment and its inhabitants.

Structural violence and climate change

The concept of structural violence is usually evoked to understand the oppression inherent in human hierarchies. The concept of structural violence can be used to illuminate the exploitation of animals and nature, as well as how environmental degradation aggravates conditions of inequality for human animals. Johan Galtung is commonly referenced for his broad definition of direct, structural, and cultural violence (Galtung, 1969, 1990). Through these conceptualizations, Galtung (1969, 1990) teases out multiple forms of violence. Direct violence refers to physical or psychological violence, acted out by one person against another, including the threat of such action. Structural violence, on the other hand, points to manifestations of injustice which are engineered into our systemic social apparatus or various institutions which inflicts harm by denying certain groups from the right to safety or the ability to meet fundamental human needs. Structural violence—against both human and nonhuman animals—is often both formally bureaucratized and institutionalized. Structural violence "shows up as unequal power and consequently, unequal life chances" (Galtung, 1969, p. 171). Finally, Galtung (1990) defines cultural violence as "any aspect of a culture to legitimize violence in its direct or structural form" (p. 291).

The structural violence of climate change manifests itself in a variety of ways. The most privileged and wealthy citizens benefit the most from environmental degradation, while simultaneously making the greatest contribution to climate change. Meanwhile, the least privileged citizens suffer the worst consequences of climate change while contributing the least to the creation of the problem (Cardenas, 2016). U.S. citizens emit, on average, 17 tons of carbon dioxide each year, while an average global citizen is responsible for emitting just 5 tons (O'Brien, 2017). Carbon dioxide accounts for about 82% of GHG emissions (Schwägerl, 2014). When carbon dioxide emission is broken down by social class, the poorest U.S. citizen emit less than 5 tons annually, while the richest are responsible for 70 tons per year (O'Brien, 2017). Approximately 80% of the world's population resides in the Global South (O'Brien, 2017). This 80% of the global population is responsible for just 20% of global GHG emissions. Meanwhile, the remaining 20% of the population residing in the Global North is responsible for 80% of GHG (Hartmann, 2009). And as scientists anticipate consequences of climate change—particularly in the form of climate-related disasters—it is clear that those most responsible for climate change will be the least affected by it (Schwägerl, 2014).

In addition to the direct consequences of the climate crisis, environmental laws may be designed to disadvantage the most vulnerable groups—yet another form of structural violence. People of color within the environmental justice movement point to examples such as the widespread and long-standing poisoning of Native America lands, Louisiana's "Cancer Alley" and the toxic air that surrounds it, as well as the botched cleanup of the weapons manufacturing site in Hanford, Washington (Verchick, 2004). These examples suggest that environmental policy, procedure, and intervention may be engineered to benefit the most powerful groups, while sacrificing the most marginal among us—both human and nonhuman.

The structural violence of climate change manifests along the lines of gender and race as well. A closer look reveals feminized and racialized causes and consequences of climate change, which are discussed in greater detail in this chapter. It is notable that when we look at the largest polluters at the institutional level, militaries and oil companies are two of the major contributors to climate change (Abramsky, 2010). These are entities that are patriarchal in character and structure. The scope of gendered climate injustice is wide. They include an increase in women's workloads in drought-stricken regions. In particular, women who are responsible for gathering food from ailing crops and increasingly scarce water endure extra burdens during drought (Momtaz, Asaduzzaman, & Taylor & Francis, 2018). Indigenous people, particularly those residing in Sub-Saharan Africa, are vulnerable to climate consequences, which highlights how climate injustices fall along racial lines (Connolly-Boutin & Smit, 2016). These marginalized groups have little efficacy in shaping environmental policy or even sharing a voice in those spaces devoted to solving the climate crisis. Conversely, the official discourse surrounding climate change is dominated by white males from wealthy countries (Black, 2016).

Direct violence and climate change

How do droughts and overpopulation contribute to changing forms of violence? It is widely accepted that violent conflict is one result of progressive climate change. The relationship is supported by compelling evidence (Adger et al., 2014). Take the conflict in Darfur, which began in 2003 and led to the deaths of nearly a half a million people. This conflict was determined to be a climate-related conflict by the United Nations (Faris, 2009) as global warming and the resulting desertification were one of the major contributors to violent episodes in Darfur. The Darfur disaster sparked a new body of literature and growing scholarly interest in climate-related conflict research (Levy, Sidel, & Patz, 2017). The crisis in Darfur has displaced thousands of Sudanese people who had to take up residence in camps when their villages were either destroyed or were too unsafe to inhabit. OXFAM conveyed reports of women being beaten or sexually assaulted in these camps with regularity, particularly when women ventured out to collect wood.

In addition to Darfur, another climate-related conflict zone is Syria. The devastating civil war in Syria began in 2011 and sparked an urgent humanitarian crisis, displacing more than nine million Syrians and causing nearly 400,000 deaths (Kelley et al., 2015). The Syrian war stems from complex circumstances, but human-induced climate change contributed to this political unraveling (Kelley et al., 2015). Syria experienced an extended drought from 2006 through 2009. This drought transformed nearly 60% of the land into desert. The Eastern Mediterranean is expected to continue to see drought and warming temperature (O'Brien, 2017).

Because the effects of climate change aggravate the existing wound of poverty and inequality, regions with poor governance and unsustainable agricultural and environmental policies see the most devastating social unrest. Water shortages have been linked to an increase in conflict (Gleick, 2010) and when the fundamental needs of a population are not met, political violence erupts. Since there is a strong association between environmental shocks and unrest (Salehyan & Hendrix, 2012), climate change will have the most profound impact on places already vulnerable to collective violence. Those who are already vulnerable will be even more so (Levy, Sidel, & Patz, 2017).

Disasters and gender violence

While gender violence occurs across all social classes and in every country of the world, such violence is more likely when families endure social upheaval, poverty, and war (Sanday, 1981b). Vandana Shiva (1988) shows how as ecological degradation has expanded, so too has violence against women, poverty among women and a decreased status for women relative to men. In India, where existing gender inequities are aggravated by environmental crisis, Shiva argues that "in the perspective of women engaged in survival struggles which are, simultaneously, struggles for the protection of nature, women and nature are intimately related, and their domination and liberation similarly linked" (1988, p. 47). Shiva (1988) reports that violence against women specifically related to dowry issues was highest in an area of India where the "green revolution" took place, Northwest India, and is part of the general upswing in violence in Punjab. Shiva (1988) elaborates by pointing out that violence against women stems from women's role displacement as the commodification of food production erodes away at women's work and women are seen as less productive (either real or imagined), which is accompanied by a decline in status: "The violence to nature, as symptomatized by the ecological crisis, and the violence to women, as symptomatized by their subjugation and exploitation arise from this subjugation of the feminine principle" (p. 6).

I should add that women's displaced role in food production, and the family violence that follows, is not just about the environmental crisis but also the neoliberal intrusion into food production in the Global South. Claudia Serrato (2010) and Richard Twine (2014) have argued that Western myopia erases indigenous foodways and ignores food colonization and the universalization of Western food practices via imperialism and globalization as well as native resistances to environmental degradation caused by colonization and globalization. Shiva (1988) goes on to point out a direct link between the destruction of ecologies and knowledge systems. These dual destructions result in violence against women as their prior status of being "knowers" has turned them into not knowing. As machines, chemicals, and processed foods replace their traditional foodways, women are displaced from productivity and development (what Shiva calls "maldevelopment"). The consequences are that natural resource bases for survival among communities in the Global South are destroyed.

While specifying levels of post-disaster violence remains methodologically difficult, a small number of studies and anecdotal evidence from post-disaster sites indicate that intimate partner violence, child abuse, and sexual violence increase during the aftermath of disasters (Delica, 1998; Fothergill, 1998, 1999, 2008; Jenkin & Phillips, 2008; Laudisio, 1993; Seager, 2006). Additionally, survivors of domestic violence face the challenge of seeking safety from their abusers while seeking shelter from the dangers associated with catastrophe (Fothergill, 1998; Laudisio, 1993). Additional evidence suggests increased crime may be one of the long-term effects of a disaster (Adger et al., 2014). In areas where human trafficking is already in operation, disasters may result in increased opportunities for traffickers (League of Red Cross and Red Crescent Societies, 1991). There may also be an increase in sexual exploitation where women experience added barriers to employment. For example, displaced women in Eastern Congo and Guinea may be coerced into sex in exchange for food or shelter and women in refugee camps may be subjected to sexual and domestic violence (League of Red Cross and Red Crescent Societies, 1991). According to OXFAM, women and girls make up 70% of the world's poor, positioning them to experience the worst impacts of climate disasters.

The Democratic Republic of the Congo is sometimes referred to as the rape capital of the world (United States Government Accountability Office & Yager, 2011). Sixty-five million Congolese people were displaced in human migration due to conflict largely connected to climate change (Codesria, 2015). Women, people of color, the poor, and children are impacted differently by disasters, particularly by droughts and floods. Women made up 90% of the victims of the cyclone and flood in Bangladesh in 1991. While there were multiple causes of this disproportionate outcome, they included that women waited too long to evacuate due to increased risk of sexual assault, instead hoping for male companions or relatives to escort them out of the area (Gaard, 2017). Rezaeian (2013) conducted a review of the literature from 1976 to 2011 and found "that being exposed to natural disasters such as tsunami, hurricane, earthquake and flood increased violence against women and girls" (p. 1105). Peggy Reeves Sanday's (1981a) famous work supports Rezaeian's (2013) conclusions. In her study of rape-free versus rape-prone societies, Sanday (1981b) found that rape was more likely in tribes that endured depleted food resources, migration, and other external conditions that required inhabitants to rely on men's destructive capacities. Further, where there is competition for resources, male roles, characterized by aggression, are given greater prestige.

This phenomenon may be more pervasive in developing countries, but increased violence against women is characteristic of a post-disaster recovery in the developed world as well. Indeed, after a 2011 earthquake in New Zealand, reports of domestic violence surged by 50% (PRI, 2011). Parkinson (2017) explored the ways in which disasters create a different context for domestic violence to unfold. Field reports following an Australian flood recording revealed that: "Human relations were laid bare and the strengths and weaknesses in relationships came more sharply into focus. Thus, socially isolated women became more isolated, domestic violence increased, and the core of relationships with family, friends and spouses were exposed" (Dobson, 1994, p. 12). Following the Loma Prieta earthquake, temporary requests for restraining orders rose by 50% and housing shortages in Santa Cruz were limiting women's ability to extract themselves from violent domestic situations (United Way of Santa Cruz County, 1990).

Because of existing gender inequalities, women and children may face greater consequences from ecological disasters than do men, but in different ways, according to their unique contexts and circumstances. Indeed, women and children are 14 times more likely to die in such calamities (Aguilar Revelo et al., 2009). Women suffer the worst consequences of climate change, not due to any innate qualities, but because of their social status, discrimination, and poverty—inequities manufactured through gendered social roles (Gaard, 2017). It is the intersection of disaster, dislocation, and existing gender inequalities that exacerbate vulnerability to violence (Jenkins & Phillips, 2008). Yet, as Gaard (2017) makes clear, "...(I)n statements of climate justice to date, there is no mention of the integral need for queer climate justice—although all our climates are both gendered and sexualized, simultaneously material, cultural, and ecological" (p. 125).

Women who survive climate change disasters may face an increased risk of sexual assault. Fothergill (1998, 1999, 2000) reports that taken collectively, a global review of 100 gender and disaster studies indicates a general increase in domestic violence following disasters. Following Hurricane Andrew, there was a 50% increase in domestic violence helpline calls (Laudisio, 1993). After Hurricane Katrina, rapes were reported by many survivors (Seager, 2006). Following the 1997 earthquake in Dale County, Alabama, reports of domestic violence increased by 600% in just the first four months (Wilson, Phillips, & Neal, 1998). Enarson's (1999) study of 77 Canadian and U.S. domestic violence programs also found a link between post-disaster domestic violence. The 1993 Missouri River Flood in the USA led to a 400% surge in women and children seeking shelter from an anti-violence coalition (Enarson, 2012). And yet, climate change discourse and solutions proceed as if gender has no relevance (MacGregor, 2010).

Jenkins and Phillips (2008) reviewed the literature on domestic violence post-disaster and arrived at a number of conclusions: Domestic violence will continue after disaster and may very well escalate; during a catastrophe, domestic violence providers may be overwhelmed and challenged to respond to all who need help; internal agency difficulties may arise as first responders are overwhelmed; the pre-disaster protective environment of survivors may collapse during catastrophe; the existing disaster response system may be unable to address the domestic violence problem as they may not recognize it as disaster-driven; and safety networks may be disrupted, weakened, or destroyed by disaster.

The LGBTQ and immigrant communities face unique discriminations during the aftermath of climate-related disasters. Following Hurricane Katrina, the religious right claimed that the disaster was God's punishment of homosexuality (Gaard, 2017). As Gaard (2017) points out, there were likely assaults on the LBGTQ population which went unrecorded after Katrina, particularly since LGBTQ persons are already living on the margins and that Louisiana lacks legal protections for this population. Moreover, the most disenfranchised groups around the world often end up being blamed for climate change, where population control and anti-immigration sentiment appear in strategies to address global GHG emissions (Williams, 2010). The overpopulation rhetoric (which is almost always directed at populations in the developing world) as a reduction strategy to minimize the worst effects of climate change is likely rooted in fear of climate refugees where desperate migrations of poor people will seek shelter in wealthier nations. Indeed, when it comes to the population–climate crisis argument, there are often tensions between climate justice and reproductive justice. Women are both blamed for the climate crisis (through childbearing) and are hit the hardest by the effects of climate disasters, not just by the hardships that follow but also by the surge in gender violence following the calamity (Gaard, 2017).

Cultural violence and nonhuman animal exploitation

On October 8, 2018, the Intergovernmental Panel on Climate Change (IPCC) released a report on climate change that communicated greater urgency to intervene in the climate crisis, stepping up and revising previous recommendations. This report outlined steps to limit global warming to 1.5°C. Prior to this IPCC 2018 report, the Paris agreement goal had established the limit at 2°C. In this report, the IPCC cautions readers about the coming disastrous consequences of the climate crisis if current warming trends are not addressed. The IPCC stresses the need for a drastic and rapid change in several consumption habits, one of which is to scale back the consumption of animal flesh. The IPCC report noted what is already well known—that the livestock sector is the leading cause of deforestation, biodiversity loss, a major contributor to climate change, and air and water pollution (Summary for Policymakers of the IPCC Special Report on Global Warming of 1.5°C approved by governments, 2018).

The widespread consumption of animal flesh is a key normative aspect of a culture that legitimizes violence in both direct and structural forms. Indeed, the environmental impact of the factory farming industry, and its effect on human populations, is astonishing. Waterways are polluted with animal waste (Williams, 2010). About 14.5% of GHG emissions is attributed to livestock production (Food and Agricultural Organization of the United States, 2015). This number is inclusive of all livestock on the planet and reflects the whole process of raising, slaughtering, transporting, and eating nonhuman animals—including the carbon made while humans eat and digest animal flesh foods. There is also a correlation between the expansion of factory farms and rural poverty, social disorganization, and crime. As the size of factory farms increase, so does rural poverty (Stull & Broadway, 2003). Moreover, crime rates and domestic violence are correlated with communities located near slaughterhouses (Eisnitz, 1997; Fitzgerald, Kalof & Dietz, 2009). David Nibert (2002, 2013) connects these oppressions by showing the inextricable bind of violence against human and nonhuman animals. Nibert (2013) traces theses intertwined phenomena historically, theoretically, and materially to show, as have other scholars that colonialism and neoliberal global capitalism thrive through the joint exploitation of marginalized human and nonhuman animal Others (Carey, 2011; Deckha, 2012; Gruen & Weil, 2012a; Nibert, 2003, 2013).

Hunger is one of the major consequences of animal agriculture. Since animal agriculture occurs on such a massive scale, the processing of animal biomass results in food insecurity for humans (Rojas-Downing et al., 2017). When land is appropriated for raising soybeans and grains for nonhuman animals, that land is then no longer available to produce plant-based foods for human animals (Rojas-Downing et al., 2017). In addition, the use of land for the increasing consumption of animal flesh foods can cause soil depletion, water insecurity, and the destruction of finite resources (Nibert, 2013). Haiti, as one example, has extreme deforestation, around 90% (Singer, 2019). Haiti now imports most of the food that they used to provide for themselves. Haiti also has the highest child hunger rate on western hemisphere (Singer, 2019). On top of the indirect effects of animal agriculture, humans who consume animal flesh foods and feminized protein face health risks (Levine, Suarez, & Brandhorst, 2014). The consumption of meat and its negative impact on human health has been well established (Levine, Suarez, Brandhorst, & 2014). Animal agriculture, particularly in its highly industrialized form with densely populated warehouses of animals present a number of other health risks such as death from antibiotic resistance/superbugs and food-borne illness (Ross, 2010).

Linking violence toward animals and gender violence

The factory farm industry is an intrinsically violent enterprise. It is challenging to envision the magnitude of slaughter and death of animals—in the billions—their rapid elimination endlessly repeated with a massive operation of reproductive control and breeding of replacement animal bodies. In just the USA alone, there are ten billion farm animals held captive and killed for eggs, milk, and meat each year (Agricultural Statistics Board, 2009). An annual estimated 50 million nonhuman animals are killed for human food globally (Sayers, 2014). The process of transforming nonhuman animals into food commodities involves a brutal process of leveraging animal bodies for profit. The sentience of nonhuman animals—the fact that they feel pain and desire to be free and reside in their natural habitat—is subjugated for profit in the factory farming industry. Factory farms have been described as practicing gratuitous cruelty toward animals in a climate of unending terror (Sanbonmatsu, 2011). Violence toward animals can be visited on them because they are "Othered". Their nonhuman status subjects animals to the exploitive profit apparatus where dynamics of violence against animals are deeply institutionalized.

As agriculture is a business with slim margins, consolidation, technologies of efficiency, and investing the very least are practices that spell more misery for animals. Just four mega companies control 59% of the pork market, 50% of the poultry market and 81% of the beef market (Stull & Broadway, 2003). Indeed, since the 1930s, the number of farms in America has shrunk by five million (Dimitri, Effland, & Conklin, 2005). The consequence of mega agricultural conglomerates consolidating farming into large-scale operations is that factory farmed animals experience super-exploitation and painful subjugation to the neoliberal goals of maximizing the "labor" that animals provide (Carey, 2011). The practice of "efficient" management of nonhuman animal bodies and technologies to extend these corporeal resources from animal bodies has exacerbated the already harsh conditions of continuous captivity for animals in the industry (Nibert, 2002, 2013; Torres, 2007).

Egg-laying hens may spend their whole lives standing on wire mesh floors, their feet often crippled as a result (Cooney, 2014). Inoculated with a regimen of antibiotics to control and maximize their reproductive success, while being bred into deformity, their distended bodies, large breasts, and tiny feet make performance of their natural functions impossible or very difficult (Davis, 1995). Due to the constant rubbing against the bars of their cramped cages, their feathers are lost. To prevent the pecking that results from the madness of confinement, egg-laying hens have their beaks cut off without anesthetic (Cooney, 2014). Male chicks are disposed of because they are not useful in the egg industry, thrown alive into grinders.

Pigs and chickens are denied natural habitats, fresh air, and contact with their families (Sayers, 2014). Pigs are highly sensitive, intelligent beings and chickens exist in complex social systems, which are profoundly disrupted when forced to live in confined, crowded spaces with other birds (Cooney, 2014). Female pigs are worse off than male pigs as their reproductive system is exploited. Pregnant sows are held captive in gestation crates, while their babies suckle through metal bars, or nursing cages, which are so small that they are unable to sit up (Sayers, 2014). It is rare for pigs and chickens to experience being outdoors (Sayers, 2014). Instead, most factories farmed animals struggle with basic mobility and are confined in spaces so small they cannot turn around (Sayers, 2014). When pigs are disabled, a common condition due to their lack of exercise and unnatural weight, they may be slaughtered (Taylor, 2017).

Farm-raised fish are also subjected to extreme confinement. They live in densely packed pools which are filthy, aggressive, and full of waste (Sayers, 2014). The animal agriculture apparatus depends on the exploitation of female nonhuman animal reproductive systems (Adams & Donavon, 1995; Donovan & Adams, 2007), which endangers the health of not just the health of nonhuman females but also the human animals who consume these reproductive secretions. Dairy cows are often hooked up to milking machines for up to 18 hours a day (Williams, 2010). The baby calves are taken away within a short time after birth. The mothers are bred to produce an unnaturally large volume of milk, an overproduction which leaves cows vulnerable to a host of physical problems. The control of female fertility for food production makes use of invasive technologies and an extreme manipulation of nonhuman animal bodies (Adams, 2003). Kept in a constant reproductive state, relentlessly force-fed, artificially inseminated, the bodies of female cows are aggressively wrung out for years and, once depleted, are finally sent to slaughter (Carey, 2011; Francione & Garner, 2010). This massive and extensive institutional suffering that human animals inflict on nonhuman animals is justified by the sharp distinction of animals as inferior (Adams & Donovan, 1995). Recognizing that harm directed toward animals is connected to human interpersonal violence is to expose these interconnected roots as embedded in a gender hierarchical social system.

Sunaura Taylor (2017) investigates the disability among animals in the factory farming industry and describes the violent rendering of sick and deformed bodies that arise from the toxic environment of the food processing and cruel technologies of bio-production:

> Industrially farmed animals live in such cramped, filthy, and unnatural conditions that disabilities become common, even inevitable. They are often crammed into cages with cement, wire, or metal-grated floors, covered in their own feces and kept in virtually nonstop darkness.... Farmed animals are bred to physical extremes: udders produce too much milk for a cow's body to hold, turkeys and chickens cannot bear the weight of their own giant breasts, and pigs' legs are too weak to support them. Chickens, turkeys, and ducks are also physically harmed by processes such as debeaking—done without anesthetic—which can leave them prone to serious infection and make it difficult for the birds to eat or preen themselves. And then there are bruises, abscesses, sores, broken bones, vaginal and reproductive disorders, chronic illnesses, and psychological issues that farmed animals are commonly reported to endure.... Chickens have been bred to grow twice as fast as they usually would, leaving them with bones and joints that cannot bear the weight of their massive forms. A battery hen, whose sole role is to lay eggs, produces around 250 eggs a year, far more than the sixty or so her body is meant to handle. The constant egg production combined with her complete inability to exercise make her prone to osteoporosis and broken bones.
>
> (p. 31)

Taylor (2017) also documents the horrifying violence inflicted on downed animals:

> Because a downed animal threatens the profit of the meat industry, abusive tactics may be used to get them up. Animal advocacy groups, included Mercy for Animals have released footage of downed animals being kicked, beaten, or dragged. Even though many of these animals would recover with rest, water and veterinary care, if a downed animal still won't get up, further violent tactics are used to "destroy" them. These impatient methods of "discarding" a downed animal include being hung to death by chains, bulldozes into piles and left to die or tossed into dumpsters (Humane Society of the United States, 2008). One employee of John Morrell and Co., a slaughterhouse in Sioux City, Iowa, described the killing of crippled pigs like this "The preferred method of handling a cripple at Morrell's is to beat him to death with a lead pipe before he gets into the chute. It's called 'piping'" while another said,

> If a hog can't walk, they scoop the son of a bitch up on a dead run with Bobcat (small tractor). Whupp! Right up in the air. If he stays in the bucket, he stays in. If he falls out, you run him over or pin him against the wall, finish busting the rest of his legs so he can't run any further.
>
> (Cited in Taylor, 2017, p. 32)

When a disease breaks out among densely populated nonhuman animal populations on industrial farms, mass slaughtering is the "solution". The public, visible, and routine nature of these mass killings highlight how our culture regards animals as killable, discardable, and lacking in any worth beyond their market value (Taylor, 2017). Swine flu, avian flu, mad cow, and foot and mouth disease are common news headlines. In the case of such outbreak, millions of animals may be gasses en mass with carbon dioxide (Taylor, 2017). In other cases, farmers had to resort to fire-extinguisher foam to kill off infected flocks: suffocation with water-based foam where death may take several minutes (Entis, 2015; Taylor, 2017). In another case, ten million cows, pigs, turkeys, and sheep, both adults and babies, were burned, shot, bulldozed into mass graves and then burned to dispose of the bodies (Taylor, 2017). This 2001 outbreak of foot and mouth disease in the UK left stacks of animal carcasses burning across England (Taylor, 2017). Terrified animals trying to escape traumatic deaths were seen running over each other desperate to flee (Taylor, 2017). Most of these animals did not carry the disease, which is easily treatable, preventable and not lethal to humans or animals, but were killed anyway due to the dictates of trade policies (Scully, 2003; Uhlig, 2002).

While the agricultural industry is responsible for the most gratuitous enslavement and slaughter, countless numbers of animals from a range of species are held captive, abused, tortured, and killed in industry, research labs, universities, circuses, zoos, and corporate product testing facilities (Francione, 2008). Products from make-up to pharmaceuticals to shoe and plastics include parts and ingredients derived from the bodies of nonhuman animals who experienced continuous captivity and then slaughter (Sayers, 2014). The captive abused and slaughtered animals are regarded as commodities for production or entertainment. These animals show signs of PTSD, mental illness, and psychosis caused by confinement (Taylor, 2017). Animals and other components of the nonhuman world will be subject to violence and perpetual inequality since the "other-than-human-life world" is already predetermined to have less import and value. In other words, our everyday institutions structure this inevitable exploitation (Sayers, 2014).

In these examples, we see how violence is inseparable from captivity, neglect, breeding, and profit. As the reproductive system of female animals in the food industry is violently used in service of profit, this shows the extent to which nonhuman animals are oppressed by beliefs about the exploitability of Othered bodies for human consumption. The consequences of such systems of harm reaches beyond nonhuman animal species. Animal oppression is profoundly entangled with human oppression. As detailed here, the industrialized animal economy—for consumption and recreation—is characterized by excessive oppression and enslavement of animals on an inconceivable, immense, and cruel scale. The justifications for such unnecessary violence are rooted in ideologies about human/animal difference where animals are inferior to humans for a variety of reasons— less capable, intelligent, lacking in "natural affinities". These conceptual distinctions are uncomfortably similar to justifications for racial, sexual, and gender hierarchies (Taylor, 2017).

Link between human and nonhuman animal violence

The fates of human and nonhuman animals are bound together, and so our liberation must also be yoked together. There is a concurrent need to challenge dehumanization but also claim our own animality and challenge speciesism where all animals are devalued (Taylor, 2017). Industrial civilization has exaggerated humans' domination over animals (Torres, 2007). A speciesist ideology has resulted in the belief that animals are commodities who are inherently disposable (Carey, 2011). As described in the previous section, exploitative factory farming constitutes the biggest source of animal oppression (Torres, 2007). And both human and nonhuman animals are exploited for their labor power (humans) and commodities (nonhuman animals) in the animal agriculture apparatus (Torres, 2007).

The social order of the meat industry ensures that animal executioners are some of the most marginal human animals. Factory farming is brutalizing for nonhuman animals, but also for human workers (Noske, 1997). The industrial "animal for food" apparatus, with its narrow profit margins, relies on the vulnerable employees, often undocumented workers who are the most marginal of the workforce (Warkentin, 2012). Mostly male immigrants, these slaughterhouse workers are at risk of death and injury as slaughterhouse work is among the most dangerous jobs in the world (Townsend, 2018). In addition to the hazards of the job, there are frequent immigrant raids where meat processing workers are arrested (Fitzgerald, 2007). And yet food corporations continue to recruit workers from Mexico and continue to subject both legal and illegal employees to extreme exploitation (Fitzgerald, 2007). Humans are exploited in numerous ways in the quest for "efficient" meat and dairy production. To practice killing and the horrifying work of cutting up carcasses, factory farm workers must suppress empathy and are traumatized by the pain they are obliged to inflict on animals (Parr, 2013). In addition to the psychological and spiritual devastation, factory farm workers face a number of health hazards. Slaughterhouse work moves fast and results in a high number of accidents and injuries where workers are wounded by the knives they use to carve up carcasses. (Grzywacz et al., 2006; Noske, 1997). Factory farm workers may be injured by animals who are fighting back against inflicted cruelty (Parr, 2013). Jason Hribal's book (2011), *Fear of the Animal Planet: The Hidden History of Animal Resistance*, documents numerous examples of animal courage in fighting back against their captors in violent acts of resistance. Most of these violent acts from animals were responses to abusive treatment or misery of confinement. Many of these animals will travel as far away as possible and there are many astonishing stories of pigs and cows who escaped from slaughterhouses, some making it safely to farm animal sanctuaries. These examples show not only the danger posed toward humans by exploiting animals in violent industries, but also that animals are not passive, that they have interests of their own, and that even the most terrorized animal will fight for their own life.

In a neoliberal marketplace, slaughterhouses tend to draw workers from "expendable" migrant labor pools and may be regarded as "disposable" workers (Nibert, 2013). Slaughterhouse workers are usually isolated in rural areas, very unlikely to submit complaints, to organize, and have few alternative employment options. In addition to psychological distress, low pay, and risk of physical injury, migrant slaughterhouse workers are less likely to report injuries due to fear of losing their jobs, language barriers, deportation, and lack of health care (Quandt et al., 2006). Of course, the domination of nature is a pre-capitalist notion, but neoliberal capitalism has turned the plunder of nature into society's way of life (Bookchin, 2005).

Violence against animals, interpersonal violence and war

A number of great thinkers have long suspected a link between violence against nonhuman animals and interpersonal violence and war. Mary Wollstonecraft, Margaret Mead, Immanuel Kant, Pythagoras, Mahatma Gandhi, and Thomas Aquinas made ardent cases that violence between humans was bound up with nonhuman animal abuse (Beirne, 2004). Leo Tolstoy reminded us that the terror of war can also be found in the slaughterhouse when he famously said that "as long as there are slaughterhouses there will be battlefields". Anna Kingsford believed that "universal peace is absolutely impossible to a carnivorous race" (Maitland & Hart, 1913). "A corpse is a corpse", as novelist and Nobel peace prize winner Isaac Singer indicated when he said that:

> As long as people will shed the blood of innocent creatures there can be no peace, no liberty, no harmony between people. Slaughter and justice cannot dwell together. As long as man continues to be the ruthless destroyer of lower living beings he will never know health or peace.

Pythagoras is believed to have said that "For as long as men massacre animals, they will kill each other. Indeed, he who sows the seed of murder and pain cannot reap joy and love". In 1921, Henry Salt wrote that "as long as man kills the lower races for food or sport, he will be ready to kill his own race for enmity. It is not this bloodshed, or that bloodshed, that must cease, but all needless bloodshed—all wanton infliction of pain or death upon our fellow-beings" (Hendrick, 1977).

This suspicion that violence against nonhuman animals is bound up with human violence has been explored in great detail by a number of contemporary scholars. Nibert (2013), in a sweeping global and historical analysis, compares the ways that the exploitation of nonhuman animals—as forced laborers, as food and resources, and as instruments of warfare—has bolstered widespread human-to-human violence. Nibert (2013) employs the concept of oppression to trace the historical trajectory of human relations with nonhuman animals. As Nibert (2013) argues, these interspecies relations are constituted through social institutions and processes that very much involve relations of power and domination. Nibert (2013) also exposed the profound damage that results from the diversion of land and water to support animal agriculture, the acceleration of war due to employing animal exploitatively in battle, and the violent consequences of centuries of killing animals for economic gain:

> This widespread violence and destruction engendered by such uses of large numbers of "domesecrated" animals encompasses both the violence experienced by the animals and the ways in which this harm has been entangled with related forms of violence against free-living animals and groups of devalued humans. These include invasion, conquest, extermination, displacement, repression, coerced and enslaved servitude, gender subordination and sexual exploitation, and hunger. Accompanying such violence have been deadly zoonotic diseases that have contributed to the destruction of entire cities, societies, and civilizations.
>
> (p. 5)

There is mounting evidence that cruelty to animals is bound up with violent behaviors toward other humans (Flynn, 2002; Gullone, 2012). Considering domestic violence, in particular, animal cruelty, especially toward domestic pets is not uncommon (Adams & Donovan, 1995; Gullone, 2012; McPhedran, 2009). While the exact mechanisms that link violence against animals to violence among humans are largely unexplored, Frank Ascione (2001) suggests that by accepting cruelty toward nonhuman animals, we may become desensitized to suffering in general, and our capacity to empathize with human suffering may be compromised. McPhedran (2009) also contends that a key factor possibly contributing to the associations between animal cruelty and interpersonal violence is a lack of ability to experience feelings of empathy.

There is a substantial amount of research that has uncovered a link between first witnessing aggression in others and later developing aggressive behavior (e.g., Cummings, 1987; Davies et al., 1999; Margolin & Gordis, 2000; Maughan & Cicchetti, 2002). It not surprising then that researchers have also found a relationship between animal cruelty and family violence (White & Heckenberg, 2014). The idea here is that the harm we inflict on nonhuman animals, we also inflict on each other. Gullone (2012) reviews the entirety of this research and report that findings taken as a whole estimate that between 29% and 75% of children in violent families have witnessed the animal cruelty. Further, somewhere between 10% and 57% have engaged in animal cruelty (Gullone, 2012). Gullone appeals to Albert Bandura's (1977, 1978, 1983) vicarious learning theory to explain these compelling findings. Bandura proposed that when the person modeling the behavior has a meaningful relationship with the observer (i.e. parent/child), that the observation of behavior is more likely to lead to performance of the observed behavior. Indeed, Gullone and Robertson (2008). Found evidence for the co-existence of animal-directed aggression and human-directed aggression in youth, and the observation of animal cruelty, as a pathway for the development of different aggressive behaviors.

The research demonstrating that animal cruelty is linked to human aggression has implications for the impact of participating in legalized violence against animals, such as rodeos, hunting, and fishing and the impact they have in shaping aggression. Gullone (2012, p. 48) concludes:

> Labelling certain aggressive behaviors as entertainment or sport because they are targeting certain species, and others as antisocial because they are targeting other species such as companion animals or humans, is incongruous. Essentially, cruelty is legalized in instances that provide benefit to humans, such as confined farming practices for pork production.

Adams (1996) reported anecdotal evidence from advocates for domestic violence survivors which point to a possible connection between hunting and violence against women. If we conceive of hunting as violence toward nonhuman animals, rather than sport, it follows that hunting may indeed inform other expressions of violence against humans and other animals. Straus (1991, 1994) advanced a "spillover theory" arguing that legal violence, such as hunting, may lead to illegal violence. Straus (1991, 1994) contents that there is a constellation of negative, antisocial outcomes that result from being spanked as a child. Corporeal punishment, like hunting, has been a socially legitimate practice. Flynn (2002) found a correlation between corporeal punishment and animal cruelty—the more often males were spanked by their fathers during their childhood, the more likely they were to have committed animal cruelty. Fynn (2002) in a comparative study of hunters and non-hunters found that hunters were twice as likely to engage in violence toward animals as non-hunters. Emel (1995, p. 708) elaborates on this process:

> If we are taught to believe or have "rationalized" that an animal is "vermin" and deserves to be killed, a feeling of sympathy can be suppressed or altogether replaced with hatred, rage, anger, or detachment. How we come to identify ourselves—as hunters, masters, victims, prey, apolitical, whatever—leads us to specific practices and forms of expression. And it is this human capacity to distance, background, deny, stereotype, and devalue the other that has led to the great atrocities of history (Gay, 1993). The cultivation of "alibis for aggression" in the name of progress has tangled ideological roots in various mediums of material and cultural conditions.

If the killing of animals through the practice of hunting "spills over" into other types of human-to-human violence, it follows that slaughterhouse work could lead to additional violence outside of the workplace. Fitzgerald, Kalof, and Dietz (2009) discovered a correlation between slaughterhouses and increased crime rates. These scholars attribute this finding to the "Sinclair effect", which references Upton Sinclair's observations in *The Jungle* that the brutality required from slaughterhouse workers translated to afterwork fights. It follows, then, that the job of killing animals in a slaughterhouse would result in emotional desensitization, distortion of ordinary sensibilities and negative psychological and social health effects (White & Heckenberg, 2014).

Jacques (2015) also discovered an association between location of slaughterhouses and increased crime rates in nearby communities. If slaughterhouse workers are engaging in violence as a vocation, it warrants investigation of how such violent practices impact human individuals. Beirne (2004) suggested that when authority and power influence a human, they are transformed by their "institutionalized social distance". It is this degree of distance and power, according to Beirne (2004), that may render extra-institutional violence more likely. MacNair (2002) discovered that slaughterhouse workers—individuals whose job it is to kill animals—have a diminished capacity to express empathy to fellow humans. The causes of violence are obviously complex and this link between animal cruelty and human violence has not received much attention from scholars, and yet it seems logical that the two share a common denominator. Further, the practice of a trans-species kinship logically extends to recognizing the inherent dignity of all living beings, which stands as a precursor to nonviolence.

Bierne (2004) explores evidence for what they call "the progression thesis"—whether there is a progression from violence inflicted on animals to violence later directed at other humans. Bierne (2004) points to other "groupings" of violent behavior. For example, in a household characterized by abuse, more than one type of family violence is likely to exist (Widom, 2000). If spousal abuse is present, then child abuse and neglect are more likely to occur there.

Conclusion

It is widely known that the destruction of the natural world poses ominous threats to all species. The current ecological crisis both originates from violence and through catastrophe and disaster, exacerbates additional violence. Violence against nonhuman animals is a normative aspect of a culture that legitimizes violence in both direct and structural forms. The ecological consequences of this widely endorsed killing of nonhuman animals in the factory farming industry, and its effect on human populations, is staggering. Domination over nature—where anything nonhuman is excluded from the social contract—locks us into a belief system that involves a seemingly changeless reality, where our everyday institutions structure this "inevitable" exploitation The profound suffering that human animals inflict on nonhuman animals is justified by hierarchical frameworks that regard the natural world as having lesser value and nonhuman animals as lacking inherent worth (Adams & Donovan, 1995). Identifying the connections between violence inflicted on nonhuman animals and human interpersonal violence expose their roots as embedded in a gender hierarchical social system where the "other-than-human-life world" is redetermined to have less import and value (Wolfe, 2010).

Captivity, neglect, and breeding for profit as well as the exploitation of the reproductive system of female animals' reproductive systems in the "food" industry reveals the inseparability of various kinds of violence. These violent practices, at their core, require that the oppression of nonhuman animals be supported by beliefs about the exploitability of Othered bodies for human consumption. The consequences of such systems of harm reaches beyond the nonhuman. Violence between human animals is also rooted in beliefs about difference where humans are considered superior to animals. These modes of "differencing" echoes justification for exclusion and discrimination based on racial, sexual, and gender hierarchies.

Chapter 5

Conclusion

Toward a nonviolent ecological society

This book has explored several connections between the gendered domination of the more-than-human-life world and gendered violence. This exploration has hopefully served to get us asking hard questions about our relationship with the biosphere—and all of its inhabitants—and how such an inquiry would lead us to also question patriarchy and other hierarchies as part of our intervention. In this text, I have stressed that if we are to overcome any hierarchy, including patriarchy, we must topple all of them, including speciesism; that the problem—as well as the solution—can be located in the way society is organized. Bookchin (2005) stresses that environmental problems are social issues and that the domination of nature by humans is an outgrowth of the hierarchy composed of humans over other humans. Even if we "eliminate social injustice ... we will not achieve social freedom. We may eliminate classes and exploitation, but we will not be spared from the trammels of hierarchy and domination" (p. 72). I follow Bookchin (2005) in stressing that the path of nonviolence includes establishing an ecological society.

Nonviolence cannot be realized unless there is a radical transformation where we are free of the domination that characterizes all social relations, including one with the earth (Ruether, 1975). One location to continue this work is with alliances between movements that share similar objectives. Environmental movements and feminist movements are intimately connected. They both are looking for the same outcome—transforming a worldview of domination and replacing it with an alternative value system (Kline, 2011; Ruether, 1975). This work means constantly and continually attending to the link between relationships of domination among humans, between humans, and between humans and nonhumans.

Humans possess a propensity to include the earth, nonhuman animals, and human animals in their sphere of moral concern, but certain ideological, institutional, and cultural factors may discourage us from doing so. I have stressed that if we are to overcome any hierarchy, including patriarchy, we must topple all of them, including speciesism. Environmental problems are social issues and the domination of nature by humans is an outgrowth of the hierarchy composed of humans over other humans. Even if we "eliminate social injustice ... we will not achieve social freedom. We may eliminate classes and exploitation, but we will not be spared from the trammels of hierarchy and domination" (Bookchin, 2005, p. 147).

The solutions to gender violence cannot be engineered, in a piecemeal fashion, to accommodate our existing social apparatus. Gender violence originates from a hierarchical, capitalist apparatus that regards the natural world as existing for human consumption. Because our fundamental orientation is one in which "man" dominates "nature", this rapacious ideology is projected onto other relations of domination. This broad, alternative framework views peace, nonviolence, human rights, and respect for the biosphere as interrelated. Gender violence is a contemporary ecosocial problem. Our anti-ecological society—characterized by gendered hierarchy, has produced a set of dislocations and social conflicts one of which is gender violence. The dualities of "otherness" that pervade our separation from nature are characterized by opposition, antagonism, and domination. The systemic beliefs that influence patterns of gendered violence are informed by a larger ideology that regards control and destruction of the biosphere as normative.

The focus of this inquiry has included actual, material, lived relationship with animals and the earth, the relational distance that informs so much of our human-to-human interaction and experience. Attention was given to ideology (belief) and structure (social reality), as well as symbolism and language. This inquiry was guided by ecofeminist insight. Ecofeminism as a practice supports healthier relationships with each other and with nonhuman entities. Ecofeminist praxis helps to undo the logic of domination in its multitude of forms. Ecofeminism stresses care as much as justice. Ecofeminist praxis involves working across boundaries by bringing feminist thinking to interrelated problems, revealing gendered power relations at the core of both gender violence and ecological degradation. In following ecofeminist logic, I contend that our response and intervention to gender violence requires adherence to ecological principles to restore humans' experience with nature.

One branch of the solution to gender violence, as well as a whole host of dislocations, repressions, and social conflicts not considered here, is to reverse the antipathy that we hold toward the natural world—a reversal of our anti-ecological society and our alienation from nature. Thomas Berry (1999) reminds us that the universe is not a collection of objects but a collection of subjects. The cultivation of reverence, rapport, and respect of other modes of being (human and nonhuman) is necessary for our restoration.

The prevention of gender violence occupies the same ethical ground as our obligations to the more-than-human-life world. To the extent that human problems are regarded as separate from the more-than-human-life world, we perpetuate sexist and dualist paradigms, which feed ideologies of "superior" and "inferior". These hierarchical ideologies inform our behavior in adverse ways. Since the domination of humans by other humans is bound up with the domination of nature, interventions must challenge the roots of such a system. Torres (2007) compels us to ask the question: "How did we end up in a society that lives and breathes domination? As humans, we not only oppress other humans, but we also dominate, abuse, and destroy nature, including the animals within it. How is it that we've stopped living in any kind of cooperative relationship with the natural world, and instead moved into one of abject exploitation and domination?" (p. 77).

Moving forward toward healing and nonviolence requires liberating our language, employing an ethic of care, crafting new masculinities, and generating new forms of consciousness. This movement toward violence prevention must also include a prohibition against cruelty and violence toward both humans and animals as a necessary universal moral imperative (Kim, 2007). Additionally, efforts to change the natural world through single-issue, piecemeal reforms, to fit the existing needs of a society based on consumption, exploitation, and hierarchy will only backfire. Instead, efforts must be made to transform our anti-ecological society into an ecological one (Bookchin, 2005). You can't have healthy people on a sick planet and so we must transform our idea of progress to fit within a future of a sustainable ecosystem (Berry, 1999). This work begins with challenging human exceptionalism and reinventing ourselves at the species level (Berry, 1999). This is the age of moral urgency that also begs for recognition of our kinship with all life on Earth. To acknowledge our embeddedness in nature reminds us that we are not separate from other forms of life.

In this final chapter, I suggest that the pursuit of a nonviolent world be taken up as a posthumanist project. Having identified particular constructions of masculinities as instrumental to oppression of both human and nonhuman Others, I begin the conversation about transforming anti-ecological hegemonic masculinities. Gender violence is a reaction against the feminized environment. Ecofeminism calls for a benevolent, nurturing, and protective approach to our environment. Traditional gender ideology regards such traits as feminine and equates them with weakness. Gender violence is therefore bound up with a rejection of all things feminine, including ethics of care (Holmes, 2016). It follows then that transformative masculinities, and eco-masculinities, offer promise to challenge traditional masculine characteristics like courage, bravery, and reason and reimagine masculinity in nonviolent forms.

I follow many other feminist scholars in advocating for a feminist ethics of care as an alternative to a punitive state-based intervention, which has been the favored intervention for several social justice movements, but the violence against women movement in particular. The hope that a hierarchical, militarized, masculinist state will effectively intervene in both gender violence and environmental degradation is an extension of the mythical male hero fantasy where a paternalistic authority rescues a vulnerable subject. We must rethink our alliances with the state and revise the narrative of "vulnerable victims" to "empowered agents". Finally, I address the vexing problem of framing survivors of violence as vulnerable subjects. Even as vulnerability is part of our shared human condition and central to our ethic of care practice, it is also a gendered and stigmatizing concept. It is important to walk the beam between maintaining sensitivity to our shared reality of vulnerability but also avoid constructing victims as persons in need of "saving". I conclude with some final thoughts on moving forward with the hope that this text advances conversation about the interrelatedness of anti-ecological beliefs, gender, and violence against human and nonhuman Others.

Posthumanism

Posthumanist scholars seek to question human exceptionalism with regard to knowing and being and also seek to move away from Western humanist traditions (Wolfe, 2010). The term "posthumanist" is used here to de-center the human in academic work and dismantle anthropocentric ideologies and systems of power (Potts, 2010). Posthumanism offers a perspective where humans are one species in a continuum of many. Posthumanist ecofeminisms work to intervene on the neoliberal subject and Western hegemony (Potts, 2010). A posthumanist perspective critiques the liberal humanist position and creates an ethical pluralism that is inclusive of all species. In *What Is Posthumanism?* Cary Wolfe (2010) argues that any version of the human and any attempts to retrieve the human by humanisms can only serve to reproduce normative subjectivities for human and nonhuman animal Others. Posthumanism does not mean "after the human", nor does it assume that all humans equally benefit from the markers and maintenance of their species; rather it notes the repositioning of the human in relation to existing matrices of power and the natural world (Wolfe, 2010).

According to ecophilosophers, human identities are mutually configuring along with other species and communities of life. We are, therefore, an *eco-self*, which is constituted in a mesh of relations with the rest of the biosphere (Fox, 1995; Mathews, 2017). Relationality dissolves dualism, and caring concern arises when one realizes the interconnectedness of all life. Ecofeminists have argued that we reject the structure of dualism and regard both women and men and Other groups as equally belonging to nature as well as culture (Gaard, 2004; Gruen, 1993; King, 1989; Plumwood, 1993; Warren, 1987).

Challenging and dismantling anthropocentrism also means including nonhuman animals in our research, scholarship, and theorizing. Scholars are generally reluctant to include nonhuman animals in their studies, perhaps influenced by the concern that oppression will be taken less seriously when studying the nonhuman (Taylor, 2017). Cudworth (2015), who argues for a posthumanist intervention in the sociological study of violence, challenges the omission of nonhuman animals in scholarship. In terms of understanding the origins of gender violence, we should not leave unexamined those violent practices that teach us to slice off or repress empathy and to distance ourselves from the "Other" (Emel, 1995, p. 731). Cudworth (2015) argues that this tendency toward human-exclusivity reveals only a partial picture of social reality, ignoring the "co-constitutive role that nonhuman animals play in social life" (p. 3). The posthumanist ecofeminist project of dismantling the regressive markers and maintenance of humanity and animality is entirely in line with feminist aims. Feminist scholars cannot deal with these issues separately. To the extent that we are successful in practicing inclusion of nonhuman animals, we will also be successful in accomplishing what Cudworth (2015) calls "disturbing a humancentric tendency" (p. 8). But even as we challenge anthropocentrism in scholarship, we must be careful not to exclude certain groups of people.

Speciesism rests on claims of animals being "different from" and "lesser than" humans. Animals are usually evaluated in terms of the human qualities that they lack rather than their own unique characteristics. When arguments are made for saving animals, those species most human-like are assigned the highest value (those with sophisticated communication abilities or intelligence). Animals are lumped into a single category and are measured against the category human, even though humans are animals and nonhuman species are wildly diverse. Human exceptionalism is challenged by all sorts of remarkable reports of nonhuman animals revealing abilities and complexities not previously known (Waal, 2019). The feminist project of destabilizing identity categories can be served by such challenges to destabilizing previous limiting understandings about animals.

Queer theorizing has thrown into question what is "authorized by nature" (Mortimer-Sandilands & Erickson, 2010). In the same ways, challenging human exceptionalism blurs the "line" between human animals and nonhuman animals. This is not to argue that animal rights should be based on nonhuman animals' "sameness" to humans. Animal rights scholars Peter Singer (1975) and Tom Reagan (1983) make this mistake by arguing for the extension of ethics to nonhuman animals on the basis of how much they are "like" us. Instead, nonhuman animals should be regarded as having their own intrinsic value, where differences are not erased, nor leveraged to serve the dominant class, but instead, the integrity of difference should be respected (Plumwood, 1993; Seager, 2003; Slicer, 1991; Warren, 1990, 1997). Vandana Shiva argues that "even the tiniest life form [must be] recognized as having intrinsic worth, integrity, and autonomy" (2000, p. 74).

New forms of consciousness

Climate scientists and environmentalists have warned us that unless humans change the way we inhabit this earth, we are headed for calamity and destruction (O'Brien, 2017). Anthropocentrism, the Otherness of all things nonhuman, the regard for the earth only in terms of human needs, has not only brought us to the brink of climate crisis, but it has informed a whole host of other social relations as well. This blueprint for oppression repeats itself wherever there are vectors of difference. Objectifying the natural world as an object for our consumption and idealizing power instead of recognizing ultimate vulnerability undergirds violent expression toward both human and nonhuman. Rearranging our understanding and our relationship will require that each person learn the spooky art of imagining a different future and cooperating in making it happen—over and over.

What might these new forms of consciousness look like? Western Philosopher Albert Schweitzer advocated for a radical answer to our many crises: a "reverence for life" (1969). Thoreau reminds us to remember the "wildness" within, that there is no such thing as a separation from nature. Cultivating new forms of consciousness requires a reconfiguration of our relationship to nature, a remembering that in nature cooperation is far more common than competition. Many ancient cultures respected vultures for keeping disease away, a reminder of our cooperative engagement in the project of living. Aldo Leopold and Schwartz (1987) compels us to "live as a citizen of the land, rather than its 'conqueror'". We must believe that the biosphere has inherent worth beyond the instrumental value of the more-than-human-life world. This living world has its own existence and mysteries that humans cannot ever fully know. In the absence of a full understanding of nonhuman lives, the prejudices and assumptions we hold about marginalized Others run so deep that we form narratives about how they should live (Harper, 2011). We should be suspicious of such narratives.

Drawing on these observations, we can enable a shared sense of well-being among all of Earth's inhabitants. To cultivate this reverence for life, to recognize the land as a benevolent host, to appreciate the inherent dignity of all life in the biosphere, requires not only new forms of consciousness but new forms of living and being. Such a shift in consciousness and the way we live would alter the understanding of not only nature but ourselves. Another branch of this work in transforming consciousness is to practice avoiding frames of language and metaphor that intensifies the oppressive binaries of patriarchy (Cuomo, 1998), such as "mother nature" and "rape of the earth". Gudmarsdottir (2010) writes:

> As Should we describe the earth as having a body? And if such a metaphoric body language is deemed helpful for contemporary feminist enquiry, how is that body gendered? Does the rape metaphor adequately express the condition of women and nature, or does it cast female bodies and the cosmos into a role that is too passive, too victimized, too

hopeless for the tastes of third wave feminists? If women struggle for subjectivity, dignity and power, in language, how helpful is it to land once more on the "wrong" side of the nature/culture scale?

(p. 207)

Avoiding these discursive pitfalls is no easy feat. While consciously altering our language patterns is part of this work, the transformation of language should arise out of a deeper commitment and practice of nondual values. Mathews (2017) suggests a discipline of practice similar to Daoism, which steers us away from the trappings of discourse, instead transforms consciousness rather than theorizing our way out. In order to foster other ways of knowing and being and of valuing otherness, differences among species should be cherished, protected, and embraced, rather than exploited for human dominance.

Eco-masculinities

It is possible to create, encourage, and practice masculinities that are nurturing, life-affirming alternatives to anti-ecological hegemonic masculinities in discourse, identity, gender performance and cultural narratives (Connell, 1990a; Gaard, 2017). Imagine if normative masculinity demanded environmental interventions? Teresa Requena-Pelegrí (2017) wonders if those familiar features of normative masculinity—namely violence, aggression, competitiveness, and emotional restraint could be channeled into meaningful environmental liberation projects. And this is a promising time for the transformation of masculine ways of being. We are seeing a greater variation in masculine models, the availability in meanings of masculinity and some progress in liberation from narrow strictures of conceptions of masculinity (Allister, 2004). Scott Russell Sanders (Allister, 2004) urges men to resist cruelty and waste. He asks a fundamental question of all masculine performance: Do we see constructions of masculinity opening things up for men—"of cleansing the doors of perception"—or closing them down (p. 47)? Reconstructing masculinity means reconceiving gender and allowing it to be diversely expressed. Multiple expressions of gender open up possibilities for eco-masculinities. Of course, "diverse expressions of gender" should still include some conditions to avoid anti-ecological hegemonic masculinity (Kheel, 2008). New eco-masculinities should not promote any of the "isms" of social domination (Allister, 2004; Mortimer-Sandilands & Erickson, 2010; Warren, 1990).

A transformation in masculinity means relinquishing the aspiration to conquer the wilderness. Stefan Brandt (2017) recalls an anecdote about former U.S. President Teddy Roosevelt's son Kermit and his "troubled" masculinity. In 1909, while on a hunting trip in Africa with father Teddy, the ritual seemed intended to transform Kermit "into a man" through the "transformative" violence of killing an exotic animal. Up until that time, father Teddy has been concerned that the 20 year old was inclined toward "effeminate" literary pursuits of reading and writing. Kermit did finally demonstrate the violent capacities that won his manhood when he killed a leopard, much to the relief of his father. Brandt (2017) goes on to point out that this Roosevelt anecdote closely links masculine identity with a conquest of nature, of taming what is wild. This theme of male initiation rites through killing animals in the 20th century is detailed in literature, cinema, and other common cultural messages. While the connection between hunting and masculinity has had a long historical association, this ritual performance of virility must be challenged.

Releasing and transforming this stereotype of men conquering nature allows a move toward heterogeneous complexity, alternative presentations of masculinity and even paradoxical interpretations of gender expression. When we do see men interfacing with nature, that interaction often plays out in hierarchical, hegemonic, and domineering ways (Kheel, 2008). Men may not realize their own participation in this hegemonic model nor the patriarchal dividends that they receive, which is why new forms of consciousness are critical to "waking up" from the spell of complicity. Indeed, men (and women for that matter) may not recognize the ways in which they unwittingly reproduce patriarchal social structures just by performing these masculinities that return to them structural privilege (Connell, 1986).

Ethic of care

Ecofeminists have made efforts to break from a rationalist tradition and instead embrace an ethic of care (Clement, 2018; Donovan & Adams, 1996). This approach involves a re-imagining of our relationship with the more-than-human-life world that recognizes the "particular other" and "attentive love" (Simone Weil), and an appreciation of individual experience in moral decision-making, compassion, sympathy, and feeling (Donovan, 1994; Robinson, 2011). Feminist ethics of care is an alternative to the rights-based discourse has dominated in social justice movements (Adams & Gruen, 2014). The feminist literature on an ethic of care, which was inspired by Carol Gilligan (1982), offers hope of moving toward nonviolence by adopting the practice of empathy. This approach does not have to result in re-feminizing care for animals, or essentializing women's kinship with the nonhuman, but rather an ethic of care should be advanced as a political stance, contextualized within an analysis of power relations, and the egregious exploitation of animals (Donovan, 1994; Gruen, 2004).

Care has been historically undervalued, coded as "feminine" and has been positioned in opposition to normative conceptualizations of masculinity (Larrabee, 2016). Care has largely been relegated to a gendered practice based on a dualistic assumption (Kheel, 2008). The persistence of this assumption is reinforced by gendered myths about rational men, their independence and self-sufficiency, precluding the reality of human dependence on others and vulnerability that is inevitably part of the human experience. Gilligan (1982) noted that feeling contrasts with conventional masculine thinking. The more-than-human-life world does not have the capacity for reason. To the extent that thinking is a prerequisite for living, animal others, who can feel, but cannot reason, will be subjected to continued violence. Valuing feeling, on the other hand, allows us to transcend species boundaries.

An ecofeminist approach celebrates caring (Salleh, 1997) and it is only from empathy that we can discover our kinship with the earth. A simple respect for nature will not ensure that power relations change, that vulnerable people will not still be exploited. What is needed now is developing an embodied awareness to the natural world, establishing new patterns of relating with the diverse life in our biosphere, and reasserting a moral imperative that the natural world is important to assist us in cultivating a more caring relationality.

An ethic of care embraces a variety of ways to value multiple forms of life—not just humans. It is our very differences, not in spite of them, but because of them that lends inherent worth and dignity to the myriad of life forms on this planet. All life should be presumed to be worth living and reason, language, and cognitive capacity should not be used as the yardstick of worthiness. Scholars, scientists, activists, and citizens need to revise the way we regard animals. New research is coming out constantly revealing how little we have previously understood about animals. While to some extent, animals can never be fully knowable, nor are we entitled to share in all their mysteries (this is a speciesist entitlement), in just the past few years we have learned that animals can think, feel, and do far more than we ever imagined possible (Ackerman & Burgoyne, 2016; Balcombe, 2006; Bekoff, 2007; Bekoff & Pierce, 2017; Montgomery, 2016; Waal, 2016, 2019).

Recent research revealing the diverse inner lives of other animals shows how intelligent and emotional animals are (Bekoff, 2007; Waal, 2019). Sun bears can mimic each other's facial expressions (Taylor et al., 2019). Animals play and experience joy and pleasure in diverse ways (Balcombe, 2006). Crabs both feel and remember pain and primates, our close kin, value justice, and fairness (Waal, 2019). The ears and noses of cows tell us how they're feeling (Bekoff & Pierce, 2017) and many animals grieve for their dead (Derbyshire, 2009; King, 2013). Some fish are known to have personalities, complex emotional lives, and pain responsiveness similar to humans (Foer, 2009). Once believed that fish did not feel pain, we now know that fish do indeed feel pain and that a humane death for fish is a myth (Foer, 2009). Some animals can remember faces, share laughter, categorize, pass on learned information to their young, are capable of lasting bonds and can communicate complex information (Bekoff & Pierce, 2017; Waal, 2016). The accumulation of new information is challenging us to revise our notions of what animals experience (Masson & McCarthy, 1995; Waal, 2016). While intelligence should not be a prerequisite for worth, our boundaries between humans and animals that were previously erected, are now crumbling (Francione & Garner, 2010).

Even without recent research showing the complexities of animal emotion, expression, and intelligence, we know that animals are sentient beings (Francione, 2008). They feel pleasure, pain, desire to live, can suffer, and will protect themselves when faced with danger (Nocella, 2014). The comparison of animals to humans is itself profoundly anthropocentric. If we show interest in and regard for animals only because they are "like us", we miss what is distinctly unique to their lives (Cochrane, 2010). Moreover, if we only compare animal others to humans, searching for similarities, we reinforce hierarchies where humans are still regarded as more worthy or valuable (Gruen, 2015). But a prerequisite for expressing care toward nonhuman animals is empathy. Empathy involves vicarious affective response toward the emotional experience of another creature or being. There are both cognitive and affective (emotional) components to empathy where the cognitive piece includes sharing an understanding of another's experience (Zahn-Waxler & Radke-Yarrow, 1990). Moral reasoning is, therefore, one important aspect of empathy along with expressing concern and reacting to the personal distress of another (Zahn-Waxler & Radke-Yarrow, 1990).

The practice of an ethic of care must include listening. We have to choose to hear. Author Arundhati Roy (2004) writes: "There's really no such thing as the 'voiceless'". There are only the deliberately silence, or the preferably unheard. Taylor (2017) elaborated on Roy's phrase by stating that

> Animals consistently voice preferences and ask for freedom. They speak to us every day when they cry out in pain or try to move away from our prods, electrodes, knives, and stun guns. Animals tell us constantly that they want out of their cages, that they want to be reunited with their families, or that they don't want to walk down the kill chute. Animals express themselves all the time, and many of us know it. If we didn't, factory farms and slaughterhouses would not be designed to constrain any choices an animal might have. We deliberately have to choose not to hear when the lobster bangs on the walls from inside a pot of boiling water or when the hen who is past her egg-laying prime struggles against the human hands that enclose her legs and neck. We have to choose not to recognize the preference expressed when the fish spasms and gasps for oxygen in her last few minutes alive.
>
> (p. 63)

Engaging in trans-species listening also involves recognizing the global rights of nature by transcending the assumptions of modernity and listening not only across human cultures but also across species (Wolfe, 2003).

In addition to developing our capacities for empathy and listening, we must cultivate a curiosity and sense of wonder toward the more-than-human-life world to realize our interconnection with it. Carson (1962) advocated cultivating a sense of wonder, to revel in the beauty of the natural world. She also critiqued prevailing scientific approaches that normalized destructive assaults on nature as a sort of progress. Carson (1962) combined science with caring. When we fail to recognize the inescapable truth that humans are part of nature, our actions will backfire. Carson (1962) encouraged humans to live with a humility that recognizes that our lives are embedded in nature. Plumwood (1993) argues for human–nature relationship not to take the form of self-other or superior–inferior, but rather as a relationship of mutuality that acknowledges both continuity and difference; where human and nonhuman worlds are sometimes overlapping, sometimes different—a continuum of relations, but always operating in a larger space of interdependence, always in flux and interaction. Ecofeminism calls for a benevolent, nurturing, and protective approach to our environment.

Connecting with and respecting nature

I refer once again to Peggy Reeves Sanday (1981a, 1981b) who found that when tribes are in harmony with their environment and respect nature, rape is usually absent. This insight challenges us to imagine what a society would look like that is in harmony with their environment. What would a modern society that respects nature look like? What if we were humbled by the powers of nature? What if we relinquished the need to control nature and surrenders to its tides and rhythms? When we seek out controlled experiences, the result is fairly consistent (e.g. TV, indoor entertainment). When we seek out experiences in nature, conditions can change on a whim and we have to be more flexible. The positive experiences in nature are more real and intense, but the disappointments are more frequent as well. Consider a backpacking trip versus visiting a casino or Disney Land. The same pattern can be observed with food. Synthetic, manufactured food is consistent and predictable. Fresh garden foods, any kind of crop, will vary wildly in their outcome from harvest to harvest. The extent to which we try to minimize these variations and live more and more in a controlled environment reduces our emotional pliability, where we come to expect consistency and become upset when expectations aren't met. Flowing with nature instead of destroying or drastically altering it can guide our building, planting, transport, and leisure. The San Francisco bay bridge was an engineering marvel that replaced the 1930s bridge. The first bridge built there was meant to defy nature (earthquake), the second was built to accommodate nature (flow, respect). Nature evolves slowly, yet humans innovate quickly. How might we align our living with the pace and flow of nature?

In addition to finding an acceptance with the ebb and flow of nature's movements, we can begin to consider and debate the rights of Nature. Could animals, rivers, and mountains also enjoy legal persons' rights that go beyond the Human? Just recently, Toledo Ohio voted to grant Lake Erie the same rights as a human (Lileks, 2019). In an act of reverence for nature, in my town of residence, Greensboro, NC, flight patterns were changed to accommodate a bald eagle nest/family. Yet 500 miles north of where I live, Canadian geese were exterminated in New York after the Hudson Bay plane incident (Migratory geese downed Hudson river plane, 2009). I can't help but be struck by the irony that we model our flight systems and devices after birds, colonize the sky, and exterminate birds whose flight path threatens our own. Chippewa medicine man Sun Bear said, "I do not think that the measure of a civilization is how tall its buildings of concrete are, but rather how well its people have learned to relate to their environment and fellow man". How might we prioritize our relationship with the environment over "progress"?

Insects, plants, microbes, mollusks, and all the living things that make up our interdependent ecosystem deserve justice. While our ethical stance toward a system of nonliving organisms may look different than the one we take toward a cow, the discussion of "many forms of justice" should occupy our hours. Ominous ecological changes have exposed the insufficiency of a regime of rights restricted to humans. Moreover, the perilous impact of colonization on indigenous peoples has helped spark a global movement to grant rights to indigenous peoples and respect the myriad of ways indigenous communities relate to nonhumans. The next step is to extend the same "individual human rights" to the "rights of nature" which would reach out to rivers, mountains, and nonhuman animals.

In thinking about the natural world as a living community, not just the environment, we cultivate a sense of the inherent value of the earth—the web of land, air, water, and growing things—every tiny bit of it. To respect, protect it and love it can replace our former neglect and can avoid the consequences of our prior ignorance. We can reverse the way we have damaged and exploited human and nonhuman beings and the land.

Rethinking alliances with the state

The state itself is invested in supporting and reproducing relations of domination, so we can't expect that the state will be a progressive force in healing the earth, combating climate change or respecting nature and its nonhuman inhabitants (Connell, 1990b). While participating in the political process is still important, our investment in change should start with local grassroots efforts, coalition building that avoids exclusionary divisive politics, and constructing social movements in our own communities (Faber, 1998). The invitation for the state to solve the problem of both climate change and gender violence has produced a number of problematic outcomes (Gilson, 2016). While we expect a positive intervention from the state, we often see the state operating instead as a mere instrument of political power (Cruikshank, 1999).

In the case of violence, reliance on the state—which is a hierarchical entity—may reproduce other systems of domination, including paternalism (Brown, 1995; Gruber, 2009). Moreover, if nonviolence is the ultimate goal, it is problematic for activists that the state responds to violent behavior with violent punishments. Ironically, the state acts out the same types of violence that we are attempting to prevent (Smith, 2005). Positive social transformation and social justice are not necessarily achieved with state intervention (Gruber, 2009). If the practice is peace, then reliance on a violent state apparatus becomes a moral problem.

While still honoring the court battles won, rights earned, and laws passed and still advocating for effective and progressive state support, we can move beyond hopes of the state as a protector and savior. We must continue to call out the state on those issues that it has not been held accountable for. And instead of relying upon the hope of a satisfying state response, we must focus on forming communities that will resist polluters and hold violent offenders accountable in humane ways (Smith, 2005). We must also think creatively about creating nonviolent ecological societies, and nonviolent, nonhierarchical communities. A transfer of agency to communities is the next iteration, as the state solution has failed. This work involves empowering individuals and communities to create environments where violence becomes "unthinkable" (Smith, 2005). This work means relocating citizens to a position where they can participate in decisions that impact their lives.

Nikolas Rose says that "to dominate is to ignore or to attempt to crush the capacity for action of the dominated. But to govern is to recognize that capacity for action and to adjust oneself to it" (1993, p. 4). These grassroots efforts must be led by the most marginal voices, rather than the dominant group talking amongst themselves. Those most affected by environmental degradation, by gender violence, are the voices that should be foregrounded. When working on liberation projects, we might reflect on the ways that our own hierarchical thinking impacts our efforts to save, liberate, help. We can avoid modeling state-based methods by continually scrutinizing our own motives and interventions to make sure that we are not reproducing violent systems, checking in to be certain that we are not inadvertently creating new forms of "domination disguised as liberation", and that we are not removing agency, empowerment, and voices from the margin in the name of "helping" (Abu-Lughod, 2013; Brennan, 2014).

May we continually ask Who counts? Who is left out? What are the conditions of inclusion and exclusion? Are we allowing others to speak for themselves? Are we imposing narratives on others? May we continually ask ourselves what kind of power arrangements are we invested in

(Abu-Lughod, 2013; Nordstrom, 1999)? Are we doing the work of confronting systems of exploitation, poverty, and harmful consumption? Are we working to create safe places, offering security, and respect? Will the lives of the most marginal have changed as a result of our efforts? Will sentient beings be any safer? Are we challenging state power or colluding with it? Does the urgency to address our social problems result in a meaningful change or are we just shoring up existing power relations?

While anti-violence women's movements have raised awareness of gender-based violence over the last 40 years, this success has come with a paradoxical twist. These movements have primarily relied on the punitive intervention of the state to correct, respond to, and prevent gender-based violence (Gruber, 2009; Howe, 2008). This call for state intervention strengthened the existing law and order politics at the time, the wildly popular "get tough" legislation backed by conservatives (Gottschalk, 2006; Hunnicutt, 2009). This consequential alliance between violence against women activists and "get tough" politicians contributed to regressive punitive conditions in the criminal justice system, aggravating existing systems of oppression, especially racism, which are linked to the original problem of violence against women (Davis, 2003; Gruber, 2009; INCITE! 2006). The anti-violence women's movement was enormously forward thinking at the time. Certainly, these scholars and activists couldn't have anticipated such regressive outcomes (Miller, 2004).

The invitation for the state to solve the problem of gender violence should be approached with caution. Given the number of problematic outcomes (Gilson, 2016), it is evident that the state has acted as an instrument of political power (Cruikshank, 1999) rather than the positive intervention activists hoped for. Enough evidence has been gathered at this point to no longer regard the criminal justice system with uncritical acceptance or as the right apparatus to correct the problem of gender-based violence (Munro & Scoular, 2012).

The state is a hierarchical entity which reproduces paternalism, as well as other systems of domination when it "disciplines" men for committing gender-based violence (Brown, 1995; Gruber, 2009). Moreover, the state responds to violence with violence. Ironically, by asking the state to intervene, it reproduces a very similar form of violence which we oppose (Smith, 2005). It is thus important to ask whether criminal law provides the achievement of social justice or positive social transformation that anti-violence activists and scholars seek (Gruber, 2009). In the pursuit and practice of peace, does the reliance on state intervention, which employs a violent response to violence make sense? And given the history of ineffectiveness and misappropriation of gender-based violence issues, should the state be our solution?

Miller (2004) presents this problem as the "freedom/protection paradox", which is the tension between maintaining a balance between critically examining the consequences of seeking protection from harm, while also defending the right to be safe, to enjoy bodily integrity, to be free from violence. Given the widespread acknowledgment among scholars about the deleterious impacts of the prison industrial complex, particularly directed at ethnic minorities, (Alexander, 2010), we are faced with an ethical problem of finding alternatives to incarceration when seeking solutions to gender violence (Smith et al., 2006). Further, the state has largely failed to provide solutions and safety for survivors of violence, which compels us further to seek alternative strategies for addressing gender violence (Smith et al., 2006).

This critique of state intervention and failure to achieve humane solutions to the problem of gender violence is not meant to ignore the hard-won battles for rights, laws, and protection. Instead, this critique is meant to hold the state accountable for its failures and to consider alternatives to relying upon the justice system. As a start, it may be worth focusing on forming communities that will hold violent offenders accountable (Smith, 2005). In thinking creatively about solutions to gender violence which do not rely upon the state, Smith (2005) offers up the promise of empowering communities to create environments where violence becomes "unthinkable". Finally, this alternative path might also involve rejecting a criminal justice system that invalidates the experience of survivors and instead places citizens in the active role of participating in governing and policy, where "...the view of a survivor as a rape victim in need of services is repositioned to that of a citizen able to participate in creating the policies affecting her life" (Miller, 2004, p. 28).

The challenge in addressing vulnerability

It is quite common to find the theme of vulnerability in discussions of the consequences of climate change and in the literature on gender violence. Even with the success of the #MeToo movement, survivors finding the courage to break the silence about their experiences of assault, harassment and abuse, and the emergence of a community of support around gender violence, being a survivor of abuse is still stigmatizing. This social stigma, and internalized oppression that may result, is gendered where women may disparagingly be referred to as "damaged goods" and men may feel emasculated. The #MeToo movement has likely taken an isolating experience of abuse and exposed the truth that abuse is everywhere, which most certainly is an empowering turn of events for survivors. Even with the gains of #MeToo, familiar gender stereotypes can remain stubbornly attached to narratives of abuse, women may be essentialized as incapable of protecting themselves and men may be regarded as weak. And yet, vulnerability is part of the human condition. It is the gendered narratives of helplessness that perpetuate stigmatizing constructions of abuse survivors.

Ultimately a focus on gender, rather than women helps us avoid constructing women as victims in need of saving, but instead regards them as citizens, and empowered agents who can lead environmental sustainability projects (Shiva, 1988) and anti-violence interventions. Gender, after all, is a system that has shaped our economic and material arrangements in such a way to produce "victims" (MacGregor, 2010). This tension around "victim politics" focuses in particular as casting women as vulnerable subjects. Vulnerability to harm is a central concern with both anti-violence work and environmentalism. Butler (2004) reminds us that vulnerability is a key condition of our shared humanity, universally experienced, and rooted in our corporeal realities. Yet, policymakers may make use of victim-centered narratives as a discursive strategy to advance their own agenda (Hyndman, 2004; Munro & Scoular, 2012). Indeed, state actors may engage "victim talk" in ways that support the aims of neoliberalism (Stringer, 2014).

Vulnerability narratives or "victim-speak" have wide cultural purchase. Vulnerability narratives often involve formula stories where pure evil and pure innocence are at odds (Loseke, 2001). Survivors of gender-based violence may participate in reproducing these familiar scripts when they tell their own story. Telling a story about a personal experience of violence is a politically formed practice, such stories are subsequently scrutinized and evaluated, and survivors may be judged on whether or not they are "credible victims" (Jobe, 2008). Jobe (2008) points out that the experiences of formerly trafficked women were compared to the stories of other trafficked women. These experiences were subject to comparison with a privileged narrative of sex trafficking. If officials validated their stories, formerly trafficked women were considered more credible.

Vulnerability is a "vexing" concept, particularly surrounding sexual violence because this term conjures up notions of dependency, weakness, and femininity (Gilson, 2016). Moreover, this feminized concept is easily exploited to serve political objectives (Gilson, 2016). Scholars have debated the sticky territory of casting survivors of gender violence as either "helpless victims" or "empowered agents" (Creek & Dunn, 2011). While vulnerability is both a "grossly under-theorized" and "ambiguous" concept (Fineman, 2008, p. 9), and has the potential for misuse, it also has great potential to take feminism beyond equality debates, point to our shared human condition and empower (Butler, 2004; Fineman, 2008, 2010, 2011). And yet, it remains true that the vulnerability to violence is contingent on intersecting realities of immigration status, language, religion, social isolation, nationality, race, ethnicity, and so on (Villalón, 2010).

The state, in particular, may appropriate women as "vulnerable" to violence in order to legitimize militarism and violence (Leatherman, 2011). War is a highly gendered project where gender scripts are employed to cast the state as "protector". In such a scenario, the "vulnerable woman" trope is evoked to reinforce men fighting to defend women (e.g. protecting family) both in our own country and in other lands (Leatherman, 2011). Such paternalistic liberation efforts to "save" women are not likely about truly rescuing women but more likely a tool to service a hidden political agenda (Mason, 2017). This "masculinization of the protector role" and the "feminization of the one who needs protection" (Pattinson, 2008) is a sort of morality theater. In this drama, the state plays the role of the hero, but the "liberation" project requires a gender narrative to render it legible.

The narrative of vulnerable woman in need of rescuing is foundational in production of militaristic projects (Abu-Lughod, 2013; Enloe, 2000; Leatherman, 2011; Nayak & Suchland, 2006), which are also bound up with environmental degradation and global economic production as military intervention is so often about securing natural resources (Mohanty, Riley, & Pratt, 2008). Ultimately, framing of women in dichotomous terms, which feed into scripts about helpless women, may ironically perpetuate gender violence. Anti-violence work, as well as environmentalism, should involve the dismantling of false binaries of female/male, private/public, and victim/perpetrator.

Another danger in the minefield of victim-speak is that these politically motivated narratives perpetuate a neoliberal agenda by framing victimization as an outgrowth of personal responsibility, thereby perpetuating a neoliberal agenda. In neoliberal ideology, "personal choice" might be given so much primacy that unequal social realities and oppressive social structures are elided. To evoke the idea of self-determination it is often in service of shifting the burden of action or non-action back to the individual and away from social causes (Hemmings & Kabesh, 2013). This sleight of hand renders individuals what Kimberly Hutchings (2013) calls either "choosers" or "losers".

If victimization is cast as something that can be avoided as long as one is in control of their lives and if "agency" is misused as a simplified path to safety, then victimization is viewed as a personal failure, rather than a result of pathological social conditions. We see this play out in familiar victim blaming, i.e. not "choosing" sobriety, caution, the right associations, clothing or destination. In this case, the causes of violence are "individualized", and gender violence once again becomes trivialized, ignored, or dismissed. Moreover, political change is avoided as long as victimization is lodged within personal responsibility or personal choice. And so "victim talk" must be approached with great care. The discussion of "vulnerable" subjects can quickly become problematic if governing bodies employ such neoliberal discourse to make invisible those social inequalities and structural disorders that aggravate violence in the first place (Rose, 1993).

Moving forward

James Nash (Gibson, 2004) reminds us that there can be no social justice without ecological justice. I would add that there can be no ecological justice without species justice. Nash (Gibson, 2004) also claims that there can be no peace among nations in the absence of peace with nature. I would add that there can be no peace among humans in the absence of peace with nature. Both climate justice and gender violence call us to understand interlocking issues of speciesism, racism, sexism, and economic injustice and other systems of inequality. I argue that it is impossible to fully understand the dynamics of either type of exploitation without considering their gendered dimensions, as well as the creation of new forms of consciousness. Moreover, seeking solutions to contemporary eco-social problems should involve drawing on insight from eco-justice philosophy, movements, and activism (Gaard, 2014).

An early impetus for the ecofeminist movement was the realization that the liberation of women and marginalized groups—the aim of all branches of feminism—cannot be fully affected without the liberation of nature, and, conversely, the liberation of nature so ardently desired by environmentalists will not be fully affected without the liberation of women. And these entangled liberatory efforts involve conceptual, symbolic, empirical, and historical linkages between women and nature as they are constructed in Western culture. Every aspect requires feminists and environmentalists to address these liberatory efforts together if we are to be successful (Warren, 1990). To date, ecofeminist theory has blossomed beyond just women and nature, exploring the connections among many issues: racism, environmental degradation, economics, electoral politics, animal liberation, reproductive politics, biotechnology, bioregionalism, spirituality, holistic health practices, sustainable agriculture, and others (Alaimo, 2000). Ecofeminist activists have worked in the environmental justice movement, the Green movement, the anti-toxins movement, the women's spirituality movement, the animal liberation movement, and the movement for economic justice (Sandler & Pezzullo, 2007). To continue to build on these efforts toward coalition, ecofeminism and anti-violence work makes sense.

Ecological impacts are experienced on bodies. Violence is experienced on bodies. If we consider this, how might we reconnect our embodied experience of inhabiting the natural world in relationship with humans, more than humans, and Other than human life? Can we explore places of environmental destruction and the violence that manifests in those spaces? In addition, what kind of gender violence happens in ecotones—those boundaries zones where the borderlands of species cross? How do climate-driven droughts and climate refugee diasporas contribute to changing forms of gender violence? How has the nexus of climate crisis, globalization, and rapid urbanization impacted gender violence? And how are the climate-related consequences of gender violence distributed among women, people of color, the economically marginal, and children? And considering all of these dimensions, how might gender violence be reduced by efforts from a transnational, intersectional feminist environmental movement? Moving forward, let us explore creative avenues of trans-species listening. We can listen to animals—in all of their varied mechanisms of communication. We can prioritize listening over "speaking for" survivors and nonhuman Others. Indeed, to speak for animals, or to speak for survivors of violence, reinforces anthropocentric and paternalistic paradigms.

Moving forward, let us strive to move away from casting women and marginalized Others as vulnerable victims in favor of a more complicated understanding of the social construction of vulnerability. While we work to uncover hidden sexism in dominating reactions against a feminized nature, let us strive to understand women's relation to nature in a non-essentializing way. Let us create spaces in nature for nonviolent gender interaction. And as we craft interventions, let us reflexively ask ourselves if our anti-violence strategies are effective or if they are in fact aggravating inequality. Environmental justice is fundamentally a counter-hegemonic challenge. And anti-violence work is fundamentally a counter-hegemonic challenge.

Our work is to make visible the interconnections between the environment and human behavior; to recognize that, for example, the peril of globalization and rapid urbanization is rooted in a rapacious social system predicated on domination of nature. Our work is to nurture the truth that there is more cooperation in nature than competition and continue to ponder the question: "can living in a different relationship with the earth change human behavior to be less violent?" (Mortimer-Sandilands, 1999). We must radically and creatively invest in this earth. We must question paradigms of "naturalness" of hierarchy. Let us really, truly "come down to earth" as a mechanism to access our innate wisdom, the best parts of ourselves.

References

Abramsky, K. (2010). *Sparking a worldwide energy revolution: Social struggles in the transition to a post-petrol world.* Oakland, CA: AK Press.

Abu-Lughod, L. (2013). *Do muslim women need saving?* (Rights, Action, and Social Responsibility). Cambridge, MA: Harvard University Press.

Ackerman, J., & Burgoyne, J. (Illustrator). (2016). *The genius of birds.* New York: Penguin Press.

Adams, C. (1990). *The sexual politics of meat: A feminist-vegetarian critical theory.* New York: Continuum International Publishing Group Inc.

Adams, C. (1993). *Ecofeminism and the sacred.* New York: Continuum.

Adams, C. (1994). *Neither man nor beast: Feminism and the defense of animals.* New York: Continuum International Publishing Group.

Adams, C. (1996). Bringing peace home: A feminist philosophical perspective on the abuse of women, children and pets. In K. Warren & D. Cady (Eds.), *Bringing peace home: Feminism, violence, and nature* (A Hypatia book) (pp. 69–87). Bloomington: Indiana University Press.

Adams, C. (2003). *The pornography of meat.* New York: Continuum.

Adams, C. (2010). Why feminist-vegan now? *Feminism and Psychology, 20*(3), 302–317.

Adams, C. (2011). *The sexual politics of meat: A feminist-vegetarian critical theory* (20th Anniversary ed.). New York: Continuum International Publishing Group.

Adams, C., & Donovan, J. (1995). *Animals and women: Feminist theoretical explorations.* Durham, NC: Duke University Press.

Adams, C., & Gruen, L. (2014). *Ecofeminism: Feminist intersections with other animals and the earth.* New York: Bloomsbury.

Adamson, J., Evans, M., & Stein, R. (2002). *The environmental justice reader: Politics, poetics, and pedagogy.* Tucson: University of Arizona Press.

Adger, W.N., Pulhin, J.M., Barnett, J., Dabelko, G.D., Hovelsrud, G.K., Levy, M., ... Vogel, C.H. (2014). Human security. In C.B Field, V.R. Barros, D.J. Dokken, K.J. Mach, M.D. Mastrandrea, T.E. Bilir, ... L.L. White (Eds.), *Climate Change 2014: Impacts, adaptation, and vulnerability. Part A: Global and sectoral aspects. Contribution of Working Group II to the Fifth Assessment Report of the Intergovernmental Panel on Climate Change* (pp. 755–791). Cambridge and New York: Cambridge University Press.

Adrinkrah, M. (2001). Patriarchal family ideology and female homicide victimization in Fiji. *Journal of Comparative Family Studies*, 32, 283–301.

Agricultural Statistics Board. (2009). *Natural agricultural statistics services and U.S. Department of Agriculture, Livestock Slaughter 2009 Summary*. Retrieved October 10, 2018, from https://usda.library.cornell.edu/concern/publications/r207tp32d?locale=en

Aguilar Revelo, L., IUCN—The World Conservation Union, United Nations Development Programme, & Global Gender and Climate Alliance. (2009). *Training manual on gender and climate change*. Gland: IUCN.

Alaimo, S. (2000). *Undomesticated ground: Recasting nature as feminist space*. Ithaca, NY: Cornell University Press.

Aldama, A.J. (2003). *Violence and the body: Race, gender, and the state*. Bloomington: Indiana University Press.

Alexander, M. (2010). *The new jim crow: Mass incarceration in the age of colorblindness*. New York: New Press.

Allister, M. (2004). *Eco-man: New perspectives on masculinity and nature* (Under the sign of nature). Charlottesville: University of Virginia Press.

American Bar Association. (2013). *Domestic violence civil protection orders (CPOs) by state*.

Animal Welfare Institute. (2018). Retrieved from https://awionline.org/content/crush-videos

Ascione, F.R. (2001). The abuse of animals and human interpersonal violence: Making the connection. In P. Arkow & F. Ascione, F. (Eds.), *Child abuse, domestic violence and animal abuse: Linking the circles of compassion for prevention and intervention* (pp. 32–54). West Lafayette, IN: Purdue University Press.

Ashcroft, B. (2013). *Postcolonial studies: The key concepts* (3rd ed.). New York: Routledge.

Austin, R.L., & Young, K.S. (2000). A cross-national examination of the relationship between gender equality and official rape rates. *International Journal of Offender Therapy and Comparative Criminology*, 44, 204–221.

Avakame, E.F. (1998). How different is violence in the home? An examination of some correlates of stranger and intimate homicide. *Criminology*, 36, 601–632.

Avakame, E.F. (1999). Females' labor force participation and intimate femicide: An empirical assessment of the backlash hypothesis. *Violence and Victims*, 14, 277–291.

Bailey, C. (2005). On the backs of animals: The valorization of reason in contemporary animal ethics. *Ethics & the Environment*, 10(1), 1–17.

Bailey, W.C. (1999). The socioeconomic status of women and patterns of forcible rape for major U.S. cities. *Sociological Focus*, 32, 43–62.

Bailey, W.C., & Peterson, R.D. (1995). Gender inequality and violence against women: The case of murder. In J. Hagan & Peterson, R.D. (Eds.), *Crime and inequality* (pp. 175–205). Palo Alto, CA: Stanford University Press.

Balcombe, J. (2006). *Pleasurable kingdom: Animals and the nature of feeling good*. London: Macmillan.

Bandura, A. (1977). *Social learning theory*. Oxford: Prentice-Hall.

Bandura, A. (1978). Social learning theory of aggression. *Journal of Communication*, Summer, 1, 12–29.

Bandura, A. (1983). Psychological mechanisms of aggression. In R.G. Geen & E.I. Donnerstein (Eds.), *Aggression: Theoretical and empirical reviews* (Vol. 1, pp. 1–40). New York: Academic Press.

Barnes, R. (2008). 'I Still Sort of Flounder Around in a Sea of Non-Language': The constraints of language and labels in women's accounts of woman-to-woman partner abuse. In K. Throsby & F. Alexander (Eds.), *Gender and interpersonal violence: language, action and representation*. New York: Palgrave Macmillan.

Baron, L., & Straus, M.A. (1987). Four theories of rape: A macrosocial analysis. *Social Problems*, 34, 467–489.

Barraclough, L. (2014). 'Horse tripping': Animal welfare laws and the production of ethnic Mexican illegality. *Ethnic and Racial Studies*, 37(11), 2110–2128. doi:10.1080/01419870.2013.800571

Bauman, Z. (2000). *Liquid modernity*. Cambridge: Polity Press.

Beauvoir, S. (1968). *The second sex*. New York: Knopf.

Beechey, V. (1979). On patriarchy. *Feminist Review*, 3, 66–82.

Beirne, P. (1999). For a nonspeciesist criminology: Animal abuse as an object of study. *Criminology*, 37(1), 117–148.

Beirne, P. (2004). From animal abuse to interhuman violence? A critical review of the progression thesis. *Society and Animals*, 12(1), 39–65.

Bekoff, M. (2007). *The emotional lives of animals: A leading scientist explores animal joy, sorrow, and empathy—and why they matter*. Novato, CA: New World Library.

Bekoff, M., & Pierce, J. (2017). *The animals' agenda: Freedom, compassion, and coexistence in the human age*. Boston, MA: Beacon Press.

Berman, T. (1994). The rape of mother nature? Women in language of environmental discourse. *Trumpeter*, 11(4), 173–178.

Berry, T. (1999). *The great work: Our way into the future* (1st ed.). New York: Bell Tower.

Birke, L. (2002). Intimate familiarities? Feminism and human-animal studies. *Society and Animals*, 10(4), 429–436.

Black, T. (2016). Race, gender, and climate injustice: Dimensions of social and environmental inequality. In P. Godfrey & D. In Torres (Eds.), *Systemic crises of global climate change: Intersections of race, class, and gender* (pp. 74–96). New York: Routledge.

Blumberg, R.L. (1979). Rural women and development: Veil of invisibility, world of work. *International Journal of Intercultural Relations*, 3, 447–472.

Blumberg, R.L. (1984). A general theory of gender stratification. *Sociological Theory*, 2, 23–101.

Boddice, R. (2011). *Anthropocentrism: Humans, animals, environments* (Human-animal studies, 1573–4226, 12). Leiden: Brill. doi:10.1163/ej.9789004187948.i-348.

Bookchin, M. (2005). *The ecology of freedom: The emergence and dissolution of hierarchy*. Oakland, CA: AK Press.

Bornstein, D. R., Fawcett, J., Sullivan, M., Senturia, K. D., & Shiu-Thornton, S. (2006). Understanding the experiences of lesbian, bisexual and trans survivors of domestic violence: A qualitative study. *Journal of Homosexuality*, 51(1), 159–181.

Bradley, H. (2013). *Gender* (2nd ed., 2nd revised and updated ed., Key concepts). Cambridge: Polity Press.

Brandt, S. (2017). The "Wild, Wild World" masculinity and the environment in the American literary imagination. In J. Armengol, M. Bosch Vilarrubias, A. Carabí, & T. Requena (Eds.), *Masculinities and literary studies: Intersections and new directions* (Routledge advances in feminist studies and intersectionality) (pp. 113–136). Milton: Taylor and Francis.

Brennan, D. (2014). *Life interrupted: Trafficking into forced labor in the United States.* Durham, NC: Duke University Press.

Brewer, V.E., & Smith, M.D. (1995). Gender inequality and rates of female homicide victimization across U.S. cities. *Journal of Research in Crime and Delinquency,* 32, 175–190.

Brown, W. (1995). *States of injury: Power and freedom in late modernity.* Princeton, NJ: Princeton University Press.

Browne, A., & Williams, K.R. (1989). Exploring the effect of resource availability and the likelihood of female-perpetrated homicides. *Law and Society Review,* 23, 75–94.

Brownmiller, S. (1975). *Against our will: Men, women and rape.* New York: Fawcett Columbine.

Butler, J. (2004). *Precarious life: The powers of mourning and violence.* London: Verso.

Butterwick, D., Lafave, M., Lau, B., & Freeman, T. (2011). Rodeo catastrophic injuries and registry: Initial retrospective and prospective report. *Clinical Journal of Sport Medicine,* 21(3), 45–59.

Cain, M., Khanam, S.R., & Nahar, S. (1979). Class, patriarchy, and women's work in Bangladesh. *Population and Development Review,* 5, 405–438.

Campbell, J. (1991). *The masks of God: Occidental mythology.* New York: Penguin Compass.

Caputi, J. (1989). The sexual politics of murder. *Gender and Society,* 3, 437–456.

Cardenas, S. (2016). A personal tale from the environmental wetback: Rethinking power, privilege, and poverty in a time of climate change politics. In P. Godfrey & D. Torres (Eds.), *Systemic crises of global climate change: Intersections of race, class, and gender.* Abingdon, Oxon: Routledge (pp. 59–81).

CARE. Canada. Gender-based violence. Retrieved January 13, 2019, from https://www.care.org/work/womens-empowerment/gender-based-violence

Carey, J. (2011). *Humane disposability: Rethinking "Food Animals," animal welfare, and vegetarianism in response to the factory farm* (Dissertation). McMaster University, Hamilton, ON.

Carrington, K. (1994). Postmodernism and feminist criminologies: Disconnecting discourses? *International Journal of the Sociology of Law,* 22, 261–277.

Carson, R. (1962). *Silent spring.* New York: Houghton Mifflin.

Cashmore, E. (2000). *Making sense of sports.* New York: Routledge.

Catalano, S. (2004). *Criminal victimization, 2003.* Washington, DC: Bureau of Justice Statistics, U.S. Department of Justice.

Chavetz, J.S. (1990). *Gender equity: An integrated theory of stability and change.* Newbury Park, CA: Sage.

Chavetz, J.S., & Dworkin, A.G. (1987). In the face of threat: Organized antifeminism in comparative perspective. *Gender and Society,* 1, 33–60.

Checker, M. (2005). *Polluted promises: Environmental racism and the search for justice in a southern town.* New York: New York University Press.

Chodorow, N. (1978). *The reproduction of mothering: Psychoanalysis and the sociology of gender*. Berkeley: University of California Press.

Chomsky, N. (1999). *Profit over People*. New York: Seven Stories Press.

Clark, E. (2012). "The Animal" and "The Feminist." *Hypatia*, 27(3), 516–520.

Clement, G. (2018). *Care, autonomy, and justice: Feminism and the ethic of care* (Feminist theory and politics). New York: Routledge.

Cochrane, A. (2010). *An introduction to animals and political theory* (The Palgrave Macmillan animal ethics series). Basingstoke: Palgrave Macmillan.

Codesria, A. (2015). *Inequality and climate change: Perspectives from the south = inégalité et changement climatique: Perspectives du sud* (Codesria book series) (G. Delgado, Ed.). Dakar: CODESRIA.

Coles, N., & Zandy, J. (2007). *American working-class literature: An anthology*. New York: Oxford University Press.

Collard, A., & Contrucci, J. (1989). *Rape of the wild: Man's violence against animals and the earth*. Bloomington: Indiana University Press.

Collins, P. (1991). *Black feminist thought: Knowledge, consciousness, and the politics of empowerment*. New York: Routledge.

Collins, P. (2000). *Black feminist thought* (Second Edition). New York: Routledge.

Collins, P. (2006). *Black sexual politics: African Americans, gender, and the new racism* (Sociology race and ethnicity). New York: Routledge.

Coltrane, S. (1998). *Gender and families*. Thousand Oaks, CA: Pine Forge Press.

Connell, R.W. (1986). *Gender and power*. Cambridge: Cambridge University Press.

Connell, R.W. (1990a). A whole new world: Remaking masculinity in the context of the environmental movement. *Gender and Society*, 4, 452–478.

Connell, R.W. (1990b). The state, gender, and sexual politics. *Theory and Society*, 19, 507–544.

Connolly-Boutin, L., & Smit, B. (2016). Climate change, food security, and livelihoods in Sub-Saharan Africa. *Regional Environmental Change*, 16(2), 385–399. doi:10.1007/s10113-015-0761-x

Cook, H.B. (1994). Cockfighting on the Venezuelan Island of Margarita: A ritualized form of male aggression. In A. Dundes (Ed.), *The cockfight: A casebook* (pp. 232–240). Madison: University of Wisconsin Press.

Cooney, N. (2014). *Veganomics: The surprising science on what motivates vegetarians, from the breakfast table to the bedroom*. Brooklyn, NY: Lantern Books.

Creek, S.J., & Dunn, J.L. (2011). Rethinking gender and violence: Agency, heterogeneity, and intersectionality. *Sociology Compass*, 5(5), 311–322.

Crenshaw, K. (1991). Mapping the margins: Intersectionality, identity politics, and violence against women of color. *Stanford Law Review*, 43 (6), 1241–1299.

Cruikshank, B. (1999). *The will to empower: Democratic citizens and other subjects*. Ithaca, NY: Cornell University Press.

Cudworth, E. (2005). *Developing ecofeminist theory: The complexity of difference*. Houndmills, Basingstoke, Hampshire: Palgrave Macmillan.

Cudworth, E. (2015). Killing animals: Sociology, species relations and institutionalized violence. *The Sociological Review*, 63(1), 1–18. doi:10.1111/1467–954X.12222

Cummings, E. (1987). Coping with background anger in early childhood (preschoolers). *Child Development*, 58(4), 976.

Cuomo, C. (1998). *Feminism and ecological communities: An ethic of flourishing.* London: Routledge.

Daems, T., & Robert, L. (2007). Crime and insecurity in liquid modern times: An interview with Zygmunt Bauman. *Contemporary Justice Review,* 10, 87–100.

Davies, P.T., Myers, R.L., Cummings, E.M., & Heindel, S. (1999). Adult conflict history and children's subsequent responses to conflict: An experimental test. *Journal of Family Psychology,* 13, 610–628.

Davis, A. (1983). *Women, race & class* (1st Vintage Books ed., Black women writers series). New York: Vintage Books.

Davis, A. (2003). *Are prisons obsolete?* New York: Seven Stories Press.

Davis, K. (1995). Thinking like a chicken: Farm animals and the feminine connection. In C. Adams & J. Donovan (Eds.), *Animals and women: Feminist theoretical explorations* (pp. 192–212). Durham, NC: Duke University Press.

Deckha, M. (2006). The salience of species difference for feminist legal theory. *Hastings Women's Law Journal,* 17(1), 1–38.

Deckha, M. (2009). Holding onto humanity: Animals dignity, and anxiety in Canada's assisted human reproduction act. *Unbound: Harvard Journal of the Legal Left,* 5(21), 21–54.

Deckha, M. (2012). Toward a postcolonial, posthumanist feminist theory: Centralizing race and culture in feminist work on nonhuman animals. *Hypatia,* 27(3), 527–545.

Deckha, M. (2013). Animal advocacy, feminism and intersectionality. *DEP.* Retrieved July 20, 2015, from http://www.unive.it/media/allegato/dep/n23-2013/Documenti/03_Deckha.pdf

Delica, Z. (1998). Women and children during disaster: Vulnerabilities and capacities. In E. Enarson & B.H. Morrow (Eds.), *The gendered terrain of disaster* (pp. 45–69). Westport, CT: Greenwood.

Derbyshire, D. (2009). Magpies grieve for their dead (and even turn up for funerals) [edition 2]. *Daily Mail,* 19, 19–19.

Derrida, J. (2008). *The animal that therefore I am* (M. Mallet, Ed.; D. Wells, Trans.). New York: Fordham University Press.

DeWees, M.A., & Parker, K.F. (2003). The political economy of urban homicide: Assessing the relative impact of gender inequality on sex-specific victimization. *Violence and Victims,* 18, 35–54.

Di Chiro, G. (2006). Teaching urban ecology: Environmental studies and the pedagogy of intersectionality. *Feminist Teacher: A Journal of the Practices, Theories, and Scholarship of Feminist Teaching,* 16(2), 98–109.

Dimitri, C., Effland, A., & Conklin, N. (2005). *The 20th century transformation of U.S. agriculture and farm policy/EIB-3 economic research service/USDA.* Retrieved July 25, 2015, from http://www.ers.usda.gov/media/259572/eib3_1_.pdf

Dinnerstein, D. (1976). *The mermaid and the Minotaur: Sex arrangements and human malaise.* New York: Harper & Row.

Dobash, R.P., & Dobash, R.E. (1979). *Violence against wives.* New York: Free Press.

Dobash, R.P., & Dobash, R.E. (1995). Reflections on findings from the violence against women survey. *Canadian Journal of Criminology,* 37, 457–484.

Dobash, R.P., Dobash, R.E., Wilson, M., & Daly, M. (1992). The myth of sexual symmetry in marital violence. *Social Problems,* 39, 71–91.

Dobson, N. (1994). From under the mud-pack: Women and the Charleville floods. *Australian Journal of Emergency Management,* 9(2): 11–13.

Dolan, C. (2014). *Into the mainstream: Addressing sexual violence against men and boys in conflict.* Workshop, 14 May. London: Overseas Development Institute.

Domestic Violence. (1989). N.C. Gen. Stat. §50B-1.

Donovan, J. (1990). Animal rights and feminist theory. *Signs,* 15(21), 350–375.

Donovan, J. (1994). *Feminist theory: The intellectual traditions of American feminism* (New expanded ed., Frederick Ungar book). New York: Continuum.

Donovan, J., & Adams, C. (1996). *Beyond animal rights: A feminist caring ethic for the treatment of animals.* New York: Continuum.

Donovan, J., & Adams, C. (2007). *The feminist care tradition in animal ethics: A reader* (J. Donovan, Ed.). New York: Columbia University Press.

Dugan, L., Nagin, D.S., & Rosenfeld, R. (1999). Explaining the decline in intimate partner homicide. *Homicide Studies,* 3, 187–214.

Dundes, A. (1994). Gallus as phallus: A psychoanalytic cross-cultural consideration of the cockfight as fowl play. In A. Dundes (Ed.), *The cockfight: A casebook* (pp. 241–282). Madison: University of Wisconsin Press.

Eaton, H., & Lorentzen, L. (2003). *Ecofeminism and globalization: Exploring culture, context and religion.* New York: Rowman & Littlefield.

Ehrenreich, B. (1976, June). What is socialist feminism? *WIN Magazine,* 3, 4–13.

Eisler, R. (1987). *The chalice and the blade: Our history, our future* (1st ed.). Cambridge, MA: Harper & Row.

Eisnitz, G. (1997). *Slaughterhouse: The shocking story of greed, neglect and inhumane treatment inside the U.S. meat industry.* Amherst, NY: Prometheus Books.

Ellis, L., & Beattie, C. (1983). The feminist explanation for rape: An empirical test. *The Journal of Sex Research,* 19, 74–93.

Emel, J. (1995). Are you man enough, big and bad enough? Ecofeminism and wolf eradication in the USA. *Environment and Planning D: Society and Space,* 13(6), 707–734. doi:10.1068/d130707

Enarson, E. (1999). Violence against women in disasters: A study of domestic violence programs in the United States and Canada. *Violence Against Women,* 5, 742–768.

Enarson, E. (2012). Women confronting natural disaster: From vulnerability to resilience. Boulder, Colo.: Lynne Rienner.

Enloe, C. (2000). *Maneuvers: The international politics of militarizing women's lives.* Berkeley: University of California Press.

Enloe, C. (2007). *Globalization and militarism: Feminists make the link.* New York: Rowman & Littlefield.

Entis, L. (2015, July 14). Will the worst bird flu outbreak in U.S. history finally make us reconsider factory farming chicken? *Theguardian.com.* Retrieved November 22, 2018.

Epstein, B. (2002). Ecofeminism and grass-roots environmentalism in the United States. In R. Hofrichter (Ed.), *Toxic struggles: The theory and practice of environmental justice* (pp. 144–152). Philadelphia, PA: New Society Publishers.

Erbaugh, B. (2007). Queering approaches to intimate partner violence. In L.L. O'Toole, J.R. Schiffman & M.L. Kiter-Edwards (Eds.), *Gender violence: Interdisciplinary perspectives* (2nd ed., pp. 451–459). New York: New York University Press.

Eriksson Baaz, M., & Stern, M. (2013). *Sexual violence as a weapon of war? Perceptions, prescriptions, problems in the Congo and beyond.* London and New York: Zed Books.

Evans, M. (2002). 'Nature' and environmental justice. In J. Adamson, M.M. Evans, & R. Stein (Eds.), *The environmental justice reader* (pp. 181–193). Tucson: University of Arizona Press.

Evans, R., Gauthier, D.K., & Forsyth, C.J. (1998). Dogfighting: Symbolic expression and validation of masculinity. *Sex Roles*, 39, 825–838.

Everhart, A., & Hunnicutt, G. (2013). Intimate partner violence among self-identified queer victims: Towards an intersectional awareness in scholarship, state-intervention, and organizing surrounding gender-based violence. In V. Demos & M.T. Segal (Eds.), *Gendered perspectives on conflict and violence: Macro and micro settings* (Advances in gender research, Vol. 18a, pp. 90–113). New York: Emerald Group Publishing.

Faber, D. (1998). *The struggle for ecological democracy: Environmental justice movements in the United States* (Democracy and ecology). New York: Guilford Press.

Faludi, S. (1999). *Stiffed: The betrayal of the American man* (1st ed.). New York: W. Morrow.

Faris, S. (2009). *Forecast: The consequences of climate change, from the amazon to the arctic, from darfur to napa valley* (1st ed.). New York: Henry Holt.

Fawcett, B. (1996). *Violence and gender relations: Theories and interventions*. London: Sage.

Fineman, M.A. (2008). The vulnerable subject: anchoring equality in the human condition. *Yale Journal of Law and Feminism*, 20(1), 1–23.

Fineman, M.A. (2010). The vulnerable subject and the responsive state. *Emory Law Journal*, 60(2), 251–275.

Fineman, M.A. (2011). Vulnerability, equality and the human condition. In Jones, J. M. (Ed.), *Gender, sexualities and law* (pp. 241–254). London: Routledge.

Firestone, S. (1972). *The dialectic of sex*. New York: Bantam.

Fitzgerald, A., Kalof, L., & Dietz, T. (2009). Slaughterhouses and increased crime rates. *Organization & Environment*, 22(2), 158–184.

Flavin, J. (2009). *Our bodies, our crimes: The policing of women's reproduction in America* (Alternative criminology series). New York: New York University Press.

Flax, J. (1993). *Disputed subjects: Essays on psychoanalysis, politics and philosophy*. New York: Routledge.

Flynn, C. (2002). Hunting and illegal violence against humans and other animals: Exploring the relationship. *Society and Animals*, 10(2), 137–154.

Foer, J. (2009). *Eating animals* (1st ed.). New York: Little, Brown and Company.

Food and Agricultural Organization of the United States. (2015). *The role of livestock in climate change*. Retrieved August 30, 2015, from http://www.fao.org/agriculture/lead/themes0/climate/en/

Fothergill, A. (1998). The neglect of gender in disaster work: An overview of the literature. In E. Enarson & B.H. Morrow (Eds.), *The gendered terrain of disaster: Through women's eyes* (pp. 11–25). Westport, CT: Praeger Publishers.

Fothergill, A. (1999). An exploratory study of woman battering in the Grand Forks flood disaster: Implications for community responses and policies. *International Journal of Mass Emergencies and Disasters*, 17, 79–98.

Fothergill, A. (2008). Domestic violence after disaster: Voices from the 1997 Grand Forks Flood. In B.D. Phillips & B.H. Morrow (Eds.), *Women and disasters: From theory to practice* (pp. 131–154). Philadelphia, PA: International Research Committee on Disasters.

Fox, B. (1993). On violent men and female victims: A comment on DeKeseredy and Kelly. *Canadian Journal of Sociology*, 18, 320–324.

Fox, W. (1995). *Toward a transpersonal ecology: Developing new foundations for environmentalism.* Albany: State University of New York Press.

Francione, G. (2008). *Animals as persons.* New York: Columbia University Press.

Francione, G., & Garner, R. (2010). *The animal rights debate abolition or regulation?* (Critical perspectives on animals: Theory, culture, science, and law). New York: Columbia University Press.

Fraser, N. (2009). Feminism, capitalism and the cunning of history. *New Left Review*, 56, 97–117.

Freeman, H. (2017). What do many lone attackers have in common? Domestic violence. *Guardian*, 28 March. Retrieved May 19, 2018, from https://www.theguardian.com/commentisfree/2017/mar/28/lone-attackers-domestic-violence-khalid-masood-westminster-attacks-terrorism

Fromm, E. (1955). *The sane society.* New York: Rinehart and Winston.

Fuchs, S. (2001). Beyond agency. *Sociological Theory*, 19, 25–40.

Gaard, G. (2001). Tools for a cross-cultural feminist ethics: Exploring ethical contexts and contents in the Makah whale hunt. *Hypatia*, 16(1), 1–26. doi:10.1111/j.1527-2001.2001.tb01046.x

Gaard, G. (2004). Toward a queer ecofeminism. In R. Stein (Ed.), *New perspectives on environmental justice: Gender, sexuality, and activism* (pp. 45–69). New Brunswick, NJ: Rutgers University Press.

Gaard, G. (2012). Speaking of animal bodies. *Hypatia*, 27(3), 520–526.

Gaard, G. (2014). Toward new ecomasculinities, ecogenders, and ecosexualities. In C. Adams & L. Gruen (Eds.), *Ecofeminism: Feminist Intersections with Other Animals and the Earth* (pp. 225–239). New York: Bloomsbury.

Gaard, G. (2017). *Critical ecofeminism* (Ecocritical theory and practice). Lanham, MD: Lexington Books.

Galtung, J. (1969). Violence, peace, and peace research. *Journal of Peace Research*, 6(3), 167–191.

Galtung, J. (1990). Cultural violence. *Journal of Peace Research*, 27(3), 291–305.

Gartner, R., & McCarthy, B. (1991). The social distribution of femicide in urban Canada, 1921–1988. *Law and Society Review*, 25, 287–308.

Gartner, R., Baker, R.K., & Pampel, F.C. (1990). Gender stratification and the gender gap in homicide victimization. *Social Problems*, 37, 593–612.

Gay, P. (1993). *The cultivation of hatred.* New York: W.W. Norton.

Geertz, C. (1973). *The interpretation of cultures.* New York: Basic Books.

Gelles, R.J. (1993). Through a sociological lens: Social structure and family violence theory. In R.J. Gelles and D.R. Loseke (Eds.), *Current controversies on family violence* (pp. 31–46). London: Sage.

Gelles, R.J., & D.R. Loseke (Eds.). (1993). *Current controversies on family violence.* London: Sage.

Gelles, R.J., Loseke, D.R., & Cavanaugh, M.M. (Eds.). (2005). *Current controversies on family violence* (2nd ed.). London: Sage.

Gibson, H. (2005). *Dog fighting detailed discussion.* Animal Legal and Historical Center, Michigan State University College of Law. Retrieved September 26, 2018, from www.animallaw.info/articles/ddusdogfighting.htm

Gibson, W. (2004). *Eco-justice—The unfinished journey.* Albany: State University of New York Press.

Gilligan, C. (1982). *In a different voice: Psychological theory and women's development.* Cambridge, MA: Harvard University Press.

Gilson, E.C. (2016). Vulnerability and victimization: Rethinking key concepts in feminist discourses on sexual violence. *Signs,* 42(1), 71.

Gleick, P. (2010). Climate change, exponential curves, water resources, and unprecedented threats to humanity (author abstract) (report). *Climatic Change,* 100(1), 125.

Goetting, A. (1991). Female victims of homicide: A portrait of their killers and the circumstances of their deaths. *Violence and Victims,* 6, 159–168.

Goldberg, J.M., & White, C. (2011). Reflections on approaches to trans anti violence education. In J.L. Ristock (Ed.), *Intimate partner violence in LGBTQ lives* (pp. 56–80). Routledge: New York.

Gottlieb, R. (2005). *Forcing the spring: The transformation of the American environmental movement* (Revised and updated ed.). Washington, DC: Island Press.

Gottschalk, M. (2006). *The prison and the gallows: The politics of mass incarceration in America.* New York: Cambridge University Press.

Gramsci, A. (1992). *Prison notebooks* (European perspectives) (J. Buttigieg, Ed.; A. Callari, Trans.). New York: Columbia University Press.

Griffin, S. (1971). Rape: The all-American crime. *Ramparts,* 2, 26–35.

Griffin, S. (1978). Woman and nature: The roaring inside her. New York: Summit Books.

Griffin, S. (1979). *Rape, the power of consciousness* (1st ed.). San Francisco, CA: Harper & Row.

Grinde, D.A., & Johansen, B.E. (1995). *Ecocide of Native America: Environmental destruction of Indian lands and peoples.* Santa Fe, NM: Clear Light.

Gruber, A. (2009). Rape, feminism, and the war on crime. *Washington Law Review,* 71(4), 581.

Gruen, L. (1993). Dismantling oppression: An analysis of the connection between women and animals. In G. Gaard (Ed.), *Ecofeminism: Women, animals, nature* (pp. 60–90). Philadelphia, PA: Temple University Press.

Gruen, L. (2004). Empathy and vegetarian commitments. In S. Sapontzis (Ed.), *Food for thought: The debate over eating meat* (pp. 90–117). Amherst, NY: Prometheus Books.

Gruen, L. (2015). *Entangled empathy: An alternative ethic for our relationships with animals.* New York: Lantern Books, a division of Booklight.

Gruen, L., & Weil, K. (2012a). Animal others. *Hypatia,* 27 (3), 477–487.

Gruen, L., & Weil, K. (2012b). Invited symposium: Feminists encountering animals. *Hypatia,* 27(3), 492–498.

Grzanka, P. (Ed.). (2014). *Intersectionality: A foundations and frontiers reader* (1st ed.). Boulder, CO: Westview Press, a member of the Perseus Books Group.

Grzywacz, J., Mari, M., Carrillo, L., Coates, M., & Burke, B. (2006). Illnesses and injuries reported by Latino poultry workers in western North Carolina. *American Journal of Industrial Medicine,* 49, 343—351.

Guadalupe-Diaz, X. (2013). An exploration of differences in the help-seeking of LGBQ victims of violence by race, economic class and gender. *Gay & Lesbian Issues & Psychology Review,* 9(1), 15–33.

Gudmarsdottir, S. (2010). Rapes of earth and grapes of wrath: Steinbeck, ecofeminism and the metaphor of rape. *Feminist Theology,* 18(2), 206–222.

Gullone, E. (2012). *Animal cruelty, antisocial behaviour, and aggression more than a link!* (Palgrave Macmillan animal ethics series). New York: Palgrave Macmillan.

Gullone, E., & Robertson, N. (2008). The relationship between bullying and animal abuse in adolescents: The importance of witnessing animal abuse. *Journal of Applied Developmental Psychology*, 29, 371–379.

Haraway, D. (1989). *Primate visions: Gender, race, and nature in the world of modern science.* New York: Routledge.

Haraway, D. (1991). *Simians, cyborgs, and women: The reinvention of nature.* New York: Routledge.

Harper, A.B. (2011). Connections: Speciesism, racism, and whiteness as the norm. In L. Kemmerer (Ed.), *Sister species: Women, animals, and social justice* (pp. 72–78). Urbana: University of Illinois Press.

Harper, A.B. (Ed.). (2010). *Sistah vegan: Black female vegans speak on food, identity, health, and society.* New York: Lantern Books.

Harrington, C. (2016). *Politicization of sexual violence: From abolitionism to peacekeeping.* London: Routledge.

Hartmann, B. (2009, Winter). 10 reasons why population control is not the solution to global warming. *DifferenTakes*, p. 57.

Harvey, D. (2005). *Neoliberalism: A brief history.* Oxford, Oxford University Press.

Hawley, F. (1993). The moral and conceptual universe of cockfighters: Symbolism and rationalization. *Society and Animals*, 1, 159–168.

Hemmings, C., & Kabesh, A.T. (2013). The feminist subject of agency: Recognition and affect in encounters with 'the Other'. In S. Madhok, A. Philips, & K. Wilson (Eds.), *Gender, agency, and coercion* (pp. 29–46). New York: Palgrave Macmillan.

Hendrick, G. (1977). *Henry salt, humanitarian reformer and man of letters.* Urbana: University of Illinois Press.

Herzog, H., Rowan, A., & Kossow, D. (2001). Social attitudes and animals. In D.J. Salem & A.N. Rowan (Eds.), *The state of the animals 2001* (pp. 55–69). Washington, DC: Humane Society Press.

Hochschartner, J. (2014). Vegan Angela Davis connects human and animal liberation. *Counterpunch*, January 24–26. Retrieved from http://www.counterpunch.org/2014/01/24/vegan-angela-davis-connects-human-and-animal-liberation/

Hochschild, A. (1999). *King Leopold's ghost: A story of greed, terror, and heroism in colonial Africa.* (First Mariner Books ed.). Boston, MA: Houghton Mifflin.

Hollander, J. (2001). Vulnerability and dangerousness: The construction of gender through conversation about violence. *Gender and Society*, 15(1), 83–109.

Holmes, C. (2016). *Ecological borderlands: Body, nature, and spirit in Chicana feminism.* Urbana: University of Illinois Press.

Homsher, D. (2001). *Women and guns: Politics and the culture of firearms in America.* Armonk, NY: M.E. Sharpe.

Hooks, B. (1981). *Ain't I a woman: Black women and feminism.* Boston, MA: South End Press.

Howe, A. (2008). *Sex, violence, and crime: Foucault and the "man" question.* New York: Routledge-Cavendish.

Hribal, J. (2011). *Fear of the animal planet: The hidden history of animal resistance.* Edinburgh: AK Press.

Humane Society of the United States. (2008). Rampant cruelty at California slaughter plant. Retrieved January 30, from httpp://www.humanesociety.org/news/news/2008/01/undercover_investigation_013008.html

Hunnicutt, G. (2009). Varieties of patriarchy and violence against women: Resurrecting "Patriarchy" as a theoretical tool. *Violence against Women*, 15(5), 553–573.

Hutchings, K. (2013). Choosers of losers? Feminist ethical and political agency in a plural and unequal world. In S. Madhok, A. Philips, & K. Wilson (Eds.), *Gender, agency, and coercion* (pp. 14–28). New York: Palgrave Macmillan.

Hyndman, J. (2004). Mind the gap: bridging feminist and political geography through geopolitics. *Political Geography*, 23(3), 307–322.

IASC (Inter-Agency Standing Committee). (2005). *Guidelines for gender-based violence interventions in humanitarian settings*. Geneva. Retrieved from https://gbvguidelines.org/wp/wp-content/uploads/2015/09/2015-IASC-Gender-based-Violence-Guidelines_lores.pdf.

Ignatieff, M. (1978). Cords of love, fetters of iron: The ideological origins of the penitentiary. In *A just measure of pain* (pp. 44–79). New York: Columbia University Press.

ILGA. (2010). Retrieved January 15, 2019, from https://ilga.org/bolivian-president-eating-estrogen-rich-chicken-makes-you-gay

INCITE!. (2006). *Color of violence: The Incite! anthology*. Cambridge, MA: South End Press.

International Indigenous Women's Forum (FIMI). (2006). *Mairin Iwanka Raya: Indigenous women stand against violence: A companion report to the United Nations secretary general's study on violence against women*. New York: FIMI, p. 28.

Irwin, P.G. (2001). Overview: The state of animal. In D.J. Salem & A.N. Rowan (Eds.), *The state of the animals 2001* (pp. 1–19). Washington, DC: Humane Society Press.

Jacobson, R. (2013). Women 'after' wars. In C. Cohn (Ed.), *Women and wars* (pp. 215–241). Cambridge, UK: Polity Press.

Jacques, J. (2015). The slaughterhouse, social disorganization, and violent crime in rural communities. *Society and Animals*, 23(6), 594–612. doi:10.1163/15685306-12341380

Jagger, A. (1983). *Feminist politics and human nature*. New York: Rowman & Littlefield.

Jasinski, J.L. (2001). Theoretical explanations for violence against women. In C.M Renzetti, J.L. Edleson, & R.K. Bergen (Eds.), *Sourcebook on violence against women* (pp. 5–22). Thousand Oaks, CA: Sage.

Jenkins, P., & Phillips, B. (2008). Battered women, catastrophe, and the context of safety after Hurricane Katrina. *NWSA Journal*, 20(3), 49–68.

Jenkins, S. (2012). Returning the ethical and political to animal studies. *Hypatia*, 27(3), 504–509.

Jobe, A. (2008). Sexual trafficking: A new story? In K. Throsby & F. Alexander (Eds.), *Gender and interpersonal Violence: Language, action and representation* (pp. 66–82). New York: Palgrave Macmillan.

Jou, C. (2017). S*upersizing urban America: How inner cities got fast food with government help*. Chicago, IL: University of Chicago Press.

Kalof, L. (2007). *Looking at animals in human history*. London: Reaktion.

Kalof, L., & Taylor, C. (2007). The discourse of dog fighting. *Humanity & Society*, 31(4), 319–333. doi:10.1177/016059760703100403

Kalof, L., Fitzgerald, A., & Baralt, L. (2004). Animals, women, and weapons: Blurred sexual boundaries in the discourse of sport hunting. *Society and Animals*, 12(3):237–251.

Kandiyoti, D. (1988). Bargaining with patriarchy. *Gender and Society*, 2, 274–289.

Keller, C. (1986). *From a broken web: Separation, sexism, and self*. Boston, MA: Beacon Press.

Kelley, C.P., Mohtadi, S., Cane, M.A., Seager, R., & Kushnir, Y. 2015. Climate change in the Fertile Crescent and implication of the recent Syrian drought. *PNAS*, 112, 3241–3246.

Kheel, M. (1995). License to kill: An ecofeminist critique of hunters' discourse. In C.J. Adams & J. Donovan (Eds.), *Animals and women: Feminist theoretical perspectives* (pp. 85–125). Durham, NC: Duke University Press.

Kheel, M. (2008). *Nature ethics*. New York: Rowman & Littlefield, pp. 35–68.

Kibria, N. (1990). Power, patriarchy, and gender conflict in the Vietnamese immigrant community. *Gender and Society*, 4, 9–24.

Kilmartin, C., & Allison, J. (2007). *Men's violence against women: Theory, research, and activism*. Mahwah, NJ: Lawrence Erlbaum Associates.

Kim, C. (2007). Multiculturalism goes imperial. *Du Bois Review*, 4(1), 233–249. doi:10.1017/S1742058X07070129

Kimmel, M. (2006). *Manhood in America* (2nd ed.). New York: Oxford University Press.

Kimmel, M. (2015). *Angry white men: American masculinity at the end of an era*. New York: Nation Books.

Kimmel, M., & Messner, M. (2004). *Men's lives* (6th ed.). Boston, MA: Pearson A and B.

King, B. (2013). *How animals grieve*. Chicago, IL: University of Chicago Press.

King, Y. (1989). The ecology of feminism and the feminism of ecology. In J. Plant (Ed.), *Healing the wounds: The promise of ecofeminism* (pp. 102–129). Philadelphia, PA: New Society Publishers.

Kline, B. (2011). *First along the river: A brief history of the U.S. environmental movement* (4th ed.). Lanham, MD: Rowman & Littlefield.

Knoepflmacher, U., & Tennyson, G. (1977). *Nature and the Victorian imagination*. Berkeley: University of California Press.

Kolodny, A. (1975). *The lay of the land: Metaphor as experience and history in American life and letters*. Chapel Hill: University of North Carolina Press

Kotz, D.M., & McDonough, T. (2010). Global neoliberalism and the contemporary social structure of accumulation. In T. McDonough, M. Reich & D.M. Kotz (Eds.), *Contemporary capitalism and its crises: Social structure of accumulation theory for the 21st century* (pp. 93–120). Cambridge: Cambridge University Press.

Kraus, C. (1993). Blue collar women & toxic-waste protest: The process of politicization. In R. Hofrichter (Ed.), *Toxic struggles: The theory and practice of environmental justice* (pp. 107–117). Philadelphia, PA: New Society Publishers.

Kuletz, V. (1998). *The tainted desert: Environmental and social ruin in the American West*. New York: Routledge.

Kwaymullina, A. (2018). Feminist ecologies: Changing environments in the anthropocene. In *You are on indigenous land: Ecofeminism, indigenous peoples and land justice* (pp. 193–208). Cham: Springer International Publishing, Palgrave Macmillan. doi:10.1007/978-3-319-64385-4_11

LaDuke, W. (1993). A society based on conquest cannot be sustained: Native peoples and the environmental crisis. In R. Hofrichter (Ed.), *Toxic struggles: The theory and practice of environmental justice* (pp. 98–106). Philadelphia, PA: New Society Publishers.

Larrabee, M. (Ed.). (2016). *An ethic of care: Feminist and interdisciplinary perspectives* (Thinking gender). London: Routledge.

Laudisio, G. (1993). Disaster aftermath: Redefining response—Hurricane Andrew's impact on I&R. *Alliance of Information and Referral Systems*, 15, 13–32.

Lawrence, E. (1982). *Rodeo, an anthropologist looks at the wild and the tame* (1st ed.). Knoxville: University of Tennessee Press.

League of Red Cross and Red Crescent Societies. (1991). *Working with women in emergency relief and rehabilitation programmes*. Field Studies Paper No. 2. Geneva.

Leatherman, J.L. (2011). *Sexual violence and armed conflict*. Cambridge: Polity Press.

Leiss, M. (1994). The domination of nature. In C. Merchant (Ed.), *Ecology* (Key concepts in critical theory) (pp. 45–69). Atlantic Highlands, NJ: Humanities Press.

Lenton, R.L. (1995). Power versus feminist theories of wife abuse. *Canadian Journal of Criminology*, 37, 305–330.

Leopold, A., & Schwartz, C. (1987). *A sand county almanac, and sketches here and there*. New York: Oxford University Press.

Levine, M., Suarez, J.A., & Brandhorst, S. (2014). Low protein intake is associated with a major reduction in IGF-1, cancer, and overall mortality in the 65 and younger but not older population. *Cell Metabolism*, 19(3), 407–417.

Levinson, D. (1989). *Family violence in cross-cultural perspective*. London: Sage.

Levy, B.S., Sidel, V.W., & J.A. Patz. (2017). Climate change and collective violence. *Annual Review of Public Health*, 38, 241–257.

Lileks, J. (2019). Citizen Erie. *National Review*, 71(5), 43–43.

Loseke, D.R. (2001). Lived realities and formula stories of "Battered Women". In J.F. Gubrium & J.A. Holstein (Eds.), *Institutional selves: Troubled Identities in a postmodern world* (pp. 107–126). Oxford: Oxford University Press.

MacGregor, S. (2010). 'Gender and climate change': From impacts to discourses. *Journal of the Indian Ocean Region*, 6(2), 223–238. doi:10.1080/19480881.2010.536669

MacGregor, S. (2017). Gender and environment: An Introduction. In S. MacGregor (Ed.), *Routledge handbook of gender and environment* (Routledge international handbooks) (pp. 1–24). Abingdon, Oxon: Routledge.

MacKenzie, J.M. (1987). The imperial pioneer and hunter and the British masculine stereotype in late Victorian and Edwardian times. In J.A. Mangan & J. Walvin (Eds.), *Manliness and morality: Middle class masculinity in Britain and America* (pp. 176–198). New York: St Martin's Press.

MacKinnon, C. (1983). Feminism, Marxism, method and the state: Toward feminist jurisprudence. *Signs*, 8, 635–658.

MacNair, R. (2002). *Perpetration-induced traumatic stress: The psychological consequences of killing* (Psychological dimensions to war and peace). London: Praeger.

Madsen, D.L. (2018). Over her dead body: Talking about violence against women in recent Chicana Writing. In S. Bahun-Radunovic (Ed.), *Violence and gender in the globalized world: The intimate and the extimate* (pp. 255–270). Milton: Taylor and Francis.

Maitland, E., & Hart, S. (1913). *Anna Kingsford, her life, letters, diary and work* (3rd ed.). London: John M. Watkins.

Margolin, G., & Gordis, E. (2000). The effects of family and community violence on children. *Annual Review of Psychology*, 51, 445–479. doi:10.1146/annurev. psych.51.1.445

Martell, L. (1994). *Ecology and society: An introduction*. Amherst: University of Massachusetts Press.

Mason, C.L. (2017). *Manufacturing urgency: The development industry and violence against women*. Saskatchewan: University of Regina Press.

Masson, J., & McCarthy, S. (1995). *When elephants weep: The emotional lives of animals*. New York: Delacorte Press.

Mathews, F. (2017). The dilemma of dualism. In S. MacGregor (Ed.), *Routledge handbook of gender and environment* (Routledge international handbooks) (pp. 54–70). Abingdon: Routledge.

Maughan, A., & Cicchetti, D. (2002). Impact of child maltreatment and inter-adult violence on children's emotion regulation abilities and socio-emotional adjustment. *Child Development*, 73, 1525–1542. doi:10.1111/1467-8624.00488

McPhedran, S. (2009). A review of the evidence for associations between empathy, violence, and animal cruelty. *Aggression and Violent Behavior*, 14(1), 1–4. doi:10.1016/j.avb.2008.07.005

Mendez, J.M. (1996). Serving gays and lesbians of color who are survivors of domestic violence. In C.M. Renzetti & C.H. Miley (Eds.), *Violence in gay and lesbian domestic partnerships* (pp. 53–59). New York: Haworth Press.

Menon, T.K. (2003). Crime against women in India: Violence within and without. *Philosophy and Social Action*, 29, 17–22.

Merchant, C. (1980). *The death of nature: Women, ecology, and the scientific revolution* (1st ed.). San Francisco, CA: Harper & Row.

Merchant, C. (1994). *Ecology* (Key concepts in critical theory). Atlantic Highlands, NJ: Humanities Press.

Merry, S.E. (2009). *Gender violence: A cultural perspective*. Malden, MA: Wiley-Blackwell.

Messerschmidt, J.W. (1993). *Masculinities and crime: Critique and reconceptualization of theory*. Lanham, MD: Rowman & Littlefield.

Messner, M.A. (2016). Forks in the road of men's gender politics: Men's rights vs feminist allies. *International Journal for Crime, Justice and Social Democracy*, 5(2), 6–20.

Mies, M., & Shiva, V. (2014). *Ecofeminism*. Atlantic Highlands, NJ: Zed Books.

Migratory geese downed Hudson river plane. (2009). *New Scientist*, 202(2712), Page 7.

Miller, A.M. (2004). Sexuality, violence against women, and human rights: Women make demands and ladies get protection. *Health and Human Rights: An International Journal*, 7(2), 16–47.

Miller, J. (2001). *One of the guys: Girls, gangs and gender*. New York: Oxford University Press.

Millet, K. (1970). *Sexual politics*. New York: Ballantine Books.

Mohanty, C. (2003). *Feminism without borders: Decolonizing theory, practicing solidarity*. Durham, NC: Duke University Press.

Mohanty, C., Riley, R., & Pratt, M. (2008). *Feminism and war: Confronting us imperialism*. London: Zed Books.

Momtaz, S., M. Asaduzzaman, & Taylor & Francis. (2018). *Climate change impacts and Women's livelihood: Vulnerability in developing countries* (1st ed., Routledge studies in hazards, disaster risk and climate change). Boca Raton, FL: Routledge.

Montgomery, S. (2016). *The soul of an octopus: A surprising exploration into the wonder of consciousness* (First Atria paperback ed.). New York: Atria Paperback.

Mooallem, J. (2010, April 4). They gay? There is a science to same-sex animal behavior. *New York Times*. Retrieved September 7, 2018, from http://www.nytimes.com/2010/04/04/magazine/04animals-t.html

Mooney, J. (2000a). *Gender, violence and the social order*. New York: St. Martin's Press.

Mooney, J. (2000b). *Gender, violence, and the social order*. Houndmills, Basingstoke, Hampshire: Macmillan Press.

Moracco, K.E., & Butts, J.D. (1998). Femicide in North Carolina, 1991–1993. *Homicide Studies*, 2, 422–446.

Morgan, W., Meier, K., Schneider, A., & Morgan, W. (2001). *Ethics in sport*. Champaign, IL: Human Kinetics.

Mortimer-Sandilands, C. (1999). *The good-natured feminist: Ecofeminism and the quest for democracy*. Minneapolis: University of Minnesota Press.

Mortimer-Sandilands, C., & Erickson, B. (2010). *Queer ecologies: Sex, nature, politics, desire*. Bloomington: Indiana University Press.

Munro, V.E., & Scoular, J. (2012). Abusing vulnerability? Contemporary law and policy responses to sex work in the UK. *Feminist Legal Studies*, 20(3), 189–206.

NAACP. Retrieved February 10, 2019, from https://www.naacp.org/history-of-lynchings/

Nayak, M., & J. Suchland (2006). Gender violence and hegemonic projects. *International Feminist Journal of Politics*, 8(4), 467–485.

NCADV. Retrieved February 11, 2019, from https://ncadv.org/blog/posts/domestic-violence-and-the-lgbtq-community

New York Times (2019, April 3). *Brunei stoning punishment for gay sex and adultery takes effect despite international outcry*. Retrieved April 9, 2019, from https://www.nytimes.com/2019/04/03/world/asia/brunei-stoning-gay-sex.html

Newman, M.L. (1999). White women's rights. New York: Oxford University Press.

Nibert, D. (2002). *Animal rights human rights: Entanglements of oppression and liberation*. Lanham, MA: Rowman & Littlefield.

Nibert, D. (2013). *Animal oppression and human violence: Domesecration, capitalism and global conflict*. New York: Columbia University Press.

Nieto-Galan, A. (2011). Antonio gramsci revisited: historians of science, intellectuals, and the struggle for hegemony. *History of Science*, 49(4), 453–478.

Nocella, A. (Ed.). (2014). *Defining critical animal studies: An intersectional social justice approach for liberation* (Counterpoints: Studies in the postmodern theory of education, vol. 448). New York: Peter Lang.

Nordstrom, C. (1999, June 1). Visible wars and invisible girls, shadow industries, and the politics of not-knowing. *International Feminist Journal of Politics*, 1, 14–33. 1461–6742.

Noske, B. (1997). *Beyond boundaries: Humans and animals.* Montreal: Black Rose Books.

O'Brien, K.J. (2017). *Violence of climate change: Lessons of resistance from nonviolent activists.* Washington, DC: Georgetown University Press.

O'Toole, L., Schiffman, J., & Edwards, M. (2007). *Gender violence: Interdisciplinary perspectives* (2nd ed.). New York: New York University Press.

Oakley, A. (1974). *The sociology of housework.* New York: Pantheon.

Oldenburg, V.T. (2002). *Dowry murder: The imperial origins of a cultural crime.* Oxford: Oxford University Press

Oliver, K. (2012). Ambivalence toward animals and the moral community. *Hypatia,* 27(3), 493–498.

Oppermann, S., & Lovino, S. (2017). *Environmental humanities: Voices from the anthropocene.* London: Rowman & Littlefield.

OXFAM. Boston, MA. Retrieved January 13, 2019, from https://www.oxfamamerica.org/explore/stories/overview-of-the-crisis-in-darfur/

Parkinson, D. (2017, March). Investigating the increase in domestic violence post disaster: An Australian case study. *Journal of Interpersonal Violence,* 34(11), 2333–2362. doi:10.1177/0886260517696876

Parr, A. (2013). *The wrath of capital: Neoliberalism and climate change politics.* New York: Columbia University Press.

Pattinson, J. (2008). 'Turning a pretty girl into a killer': Women, violence and clandestine operations during the Second World War. In K. Throsby & F. Alexander (Eds.), *Gender and interpersonal violence: Language, action and representation* (pp. 130–162). New York: Palgrave Macmillan.

Puar, J.K. (2007). *Terrorist assemblages: Homonationalism in queer times.* Durham: Duke University Press.

Pellow, D., & Brulle, R. (2005). *Power, justice, and the environment: A critical appraisal of the environmental justice movement* (Urban and industrial environments). Cambridge, MA: MIT.

Phillips, M., & Rumens, N. (2016). *Contemporary perspectives on ecofeminism.* London: Routledge.

Pinar, W. (2001). *The gender of racial politics and violence in America: Lynching, prison rape, and the crisis of masculinity* (Counterpoints: Studies in the postmodern theory of education, v. 163). New York: Peter Lang.

Pinker, S. (2011). *The better angels of our nature: Why violence has declined.* New York: Viking.

Piven, F.F., & Cloward, R.A. (1993). *Regulating the poor: The functions of public welfare.* New York: Vintage Books.

Plant, J. (1989). *Healing the wounds: The promise of ecofeminism.* Philadelphia, PA: New Society.

Plumwood, V. (1993). *Feminism and the mastery of nature.* London: Routledge.

Plumwood, V. (1996). 'Being prey', from Travellers' Tales: The ultimate journey. Retrieved from val.plumwood.files@wordpress.com

Potts, A. (2010). Introduction: Combating speciesism in psychology and feminism. *Feminism and Psychology,* 20(3), 291–301.

PRI. (2011, February 28). Domestic violence rages in NZ quake aftermath. Retrieved October 19, 2018, from https://www.pri.org/stories/2011-02-28/new-zealand-domestic-violence-surges-after-earthquake

Price, J. (2012). *Structural violence: Hidden brutality in the lives of women*. Albany: State University of New York Press.

Price, L.S. (2005). *Feminist frameworks: Building theory on violence against women*. Halifax: Fernwood.

Pridemore, W.A., & Freilich, J.D. (2005). Gender equity, traditional masculine culture, and female homicide victimization. *Journal of criminal justice*, 33, 213–223.

Quandt, S., Grzywacz, J., Marın, A., Carrillo, A., Coates, M.L., Burke, B., & Arcury, T.A. (2006). Illnesses and injuries reported by Latino poultry workers in western North Carolina. *American Journal of Industrial Medicine*, 49, 343–351.

Rajan, J. (2018). Women, violence, and the Islamic state: Resurrecting the caliphate through femicide in Iraq and Syria. In S. Bahun-Radunovic (Ed.), *Violence and gender in the globalized world: The intimate and the extimate* (pp. 45–89). Milton: Taylor and Francis.

Reagan, T. (1983). *The case for animal rights*. Berkeley: University of California Press.

Reiter, R. (Ed.). (1975). *Toward an anthropology of women*. New York: Monthly Review Press.

Rennison, C.M., & S. Welchans. (2000). *Intimate partner violence*. Washington, DC: Bureau of Justice Statistics, U.S. Department of Justice.

Renzetti, C.M. (1996) The Poverty of Services for Battered Lesbians. *Journal of Gay & Lesbian Social Services*, 4(1), 61–68, doi:10.1300/J041v04n01_07.

Renzetti, C.M., Edleson, J.L., &.Bergen, R.K. (2001). *Sourcebook on violence against women*. Thousand Oaks, CA: Sage.

Requena-Pelegrí, T. (2017). Green intersections caring masculinities and the environmental crisis. In J. Armengol, M. Bosch Vilarrubias, A. Carabí & T. Requena (Eds.), *Masculinities and literary studies: Intersections and new directions* (Routledge advances in feminist studies and intersectionality) (pp. 78–99). Milton: Taylor and Francis.

Rezaeian, M. (2013). The association between natural disasters and violence: A systematic review of the literature and a call for more epidemiological studies. *Journal of Research in Medical Sciences*, 18, 1103–1107.

Ritchie, A.J. (2006). Law enforcement violence against women of color. In Smith, A., Richie, B., Sudbury, J., & White, J. (Eds.), *Color of violence: The incite! Anthology* (pp. 138–156). New York: South End Press.

Robinson, F. (2011). *The ethics of care: A feminist approach to human security* (Global ethics and politics). Philadelphia, PA: Temple University Press.

Rocheleau, D., Thomas-Slayter, B., & Wangari, E. (1996). *Feminist political ecology: Global issues and local experiences* (International studies of women and place). London: Routledge.

Rogers, R. (2008). Beasts, burgers, and hummers: Meat and the crisis of masculinity in contemporary television advertisements. *Environmental Communication: A Journal of Nature and Culture*, 2(3), 281–301.

Rogin, M.P. (1987). Liberal society and the Indian question. In *Ronald Reagan, the movie: And other episodes in political demonology* (pp. 134–168). Berkeley: University of California Press.

Rojas-Downing, M., Nejadhashemi, A., Harrigan, T., & Woznicki, S. (2017). Climate change and livestock: Impacts, adaptation, and mitigation. *Climate Risk Management*, 16, 145–163. doi:10.1016/j.crm.2017.02.001

Rose, N. (1993). Government, authority and expertise in advanced liberalism. *Economy and Society*, 22(3): 283–299.

Ross, S. (2010). Food for thought, part I: Foodborne illness and factory farming. *Holistic Nursing Practice*, 24(3), 169–173.

Roy, A. (2004). The 2004 Sydney Peace Prize lecture Delivered by Arundhati Roy, 3 November 2004 at the Seymour Theatre Centre, University of Sydney.

Ruether, R. (1975). *New woman, new earth: Sexist ideologies and human liberation*. New York: Seabury Press.

Ruether, R. (1983). *Sexism and God-talk: Toward a feminist theology*. Boston, MA: Beacon Press.

Ruether, R. (1992). *Gaia and God: An ecofeminist theology of earth healing* (1st ed.). San Francisco, CA: Harper & Row.

Russell, D.E.H. (1975, December 4). *The politics of rape*. New York: Stein and Day.

Russo, A. (2006). The feminist majority foundation's campaign to stop gender apartheid. *International Feminist Journal of Politics*, 8, 557–580.

Ruth, P., Gilbert, P., & Eby, K. (2004). *Violence and gender: An interdisciplinary reader*. Upper Saddle River, NJ: Pearson Prentice-Hall.

Ryan, M., Walkowitz, J., & Newton, J. (Eds.). (1983). *The doubled vision: Sex and class in women's history*. New York: Routledge and Kegan Paul.

Ryder, R.D. (1989). Animal revolution: Changing attitudes towards speciesism. Oxford, NY: Oxford University Press.

Salehyan, I., & Hendrix, C.S. 2012. Climate shocks and political violence. *Presented at Annual Convention of the International Studies Association*, April 1, San Diego, CA.

Salleh, A. (1997). *Ecofeminism as politics: Nature, Marx, and the postmodern*. London: Zed Books.

Sanbonmatsu, J. (Ed.). (2011). *Critical theory and animal liberation*. Lanham, MD: Rowman & Littlefield.

Sanday, P.R. (1981a). *Female power and male dominance: On the origins of sexual inequality*. Cambridge: Cambridge University Press.

Sanday, P.R. (1981b). The socio-cultural context of rape: A cross-cultural study. *Journal of Social Issues*, 37, 5–27.

Sandler, R., & Pezzullo, P. (2007). *Environmental justice and environmentalism: The social justice challenge to the environmental movement* (Urban and industrial environments). Cambridge, MA: MIT Press.

Sapontzis, S. (2004). *Food for thought: The debate over eating meat*. Amherst, NY: Prometheus Books.

Sassoon, A.S. (1980). *Gramsci's politics*. London, Croom Helm.

Sayers, D. (2014). The most wretched of beings in the cage of capitalism. *International Journal of Historical Archaeology*, 18, 529–554.

Schechter, S. (1982). *The visions and struggles of the battered women's movement*. Boston, MA: South End Press.

Schwägerl, C. (2014). *The anthropocene: The human era and how it shapes our planet*. Santa Fe, NM: Synergetic Press.

Schweitzer, A. (1969). *Reverence for life* (1st ed.). New York: Harper & Row.

Schwendinger, J., & Schwendinger, H. (1993). Rape, sexual inequality, and levels of violence. In D.F. Greenberg (Eds), *Crime and capitalism* (2nd ed., pp. 82–114). Philadelphia, PA: Temple University Press.

Scott, J. (2010). Sociology and the sociological imagination: Reflections on interdisciplinarity and intellectual specialisation. In J. Burnett, S. Jeffers, & G. Thomas (Eds.), *New social connections: Sociology's subjects and objects*. Basingstoke: Palgrave.

Scott, J.C. (1985). *Weapons of the weak: Everyday forms of peasant resistance*. New Haven, CT: Yale University Press.

Scully, M. (2003). *Dominion: The power of man, the suffering of animals, and the call to mercy* (1st ed.). New York: St. Martin's Griffin.

Seager, J. (1993). *Earth Follies: Coming to feminist terms with the global environmental crisis*. New York: Routledge.

Seager, J. (2003). Pepperoni or Broccoli? On the cutting wedge of feminist environmentalism. *Gender, Place and Culture*, 10(2):167–174.

Seager, J. (2006). Noticing gender (or not) in disasters. *Geoforum*, 37(1), 2–3. doi:10.1016/j.geoforum.2005.10.004

Seibert, E. (2012). *The violence of scripture: Overcoming the old testament's troubling legacy* (JSTOR EBA). Minneapolis, MN: Fortress Press.

Serpell, J.A. (2001). Working out the beast: An alternative history of humanity's humaneness. In P. Arkow & F. Ascione (Eds.), *Child abuse, domestic violence and animal abuse: Linking the circles of compassion for prevention and intervention* (pp. 99–129). West Lafayette, IN: Purdue University Press.

Serrato, C. (2010). Ecological indigenous foodways and the healing of all our relations. *Journal for Critical Animal Studies*, 8(3), 52–60.

Shepherd, L.J. (2008). *Gender, violence and security*. London: Zed Books.

Shiva, V. (1988). *Staying alive: Women, ecology, and development*. London: Zed Books.

Shiva, V. (2000). *Stolen harvest: The hijacking of the global food supply*. Cambridge, MA: South End Press.

Shukin, N. (2009). *Animal capital: Rendering life in biopolitical times* (Posthumanities, 6). Minneapolis: University of Minnesota Press

Siisiäinen, L. (2019). *Foucault, biopolitics and resistance* (Interventions (Routledge) (firm)). Abingdon, Oxon: Routledge.

Singer, M. (2019). *Climate change and social inequality: The health and social costs of global warming* (Routledge advances in climate change research). London: Routledge.

Singer, P. (1975). *Animal liberation*. New York: New York Review.

Sjoberg, L. (2010). *Gender and international security: Feminist perspectives* (Routledge critical security studies series). London: Routledge.

Skilbeck, R. (2001). The shroud over Algeria: Femicide, Islamism and the Hijab. In D. Russell & R. Harmes (Eds.), *Femicide in global perspective* (pp. 74–99). New York: Teachers College Press.

Smart, C. (1989). *Feminism and the power of law*. London: Routledge.

Smith, A. (2005). *Conquest: Sexual Violence and American Indian Genocide*. Boston, MA: South End Press.

Smith, A., Richie, B., Sudbury, J., & White, J. (2006). The color of violence: Introduction. In Smith, A., Richie, B., Sudbury, J., & White, J. (Eds.), *Color of Violence: The Incite! Anthology* (pp. 1–12). Cambridge, MA: South End Press.

Smith, A., & LaDuke, W. (2015). *Conquest: Sexual violence and American Indian genocide*. Durham, NC: Duke University Press.

Smith, D.M., & Brewer, V.E. (1995). Female status and the gender gap in U.S. homicide investigation. *Violence Against Women*, 1, 339–350.

Spellman, E. (1982). Woman as body: Ancient and contemporary views. *Feminist Studies*, 8(1), 109–131. doi:10.2307/3177582

Spiegel, M. (1996). *The dreaded comparison: Human and animal slavery* (Revised and expanded ed.). New York: Mirror Books.

Spretnak, C. (Ed.). (1982). *The politics of women's spirituality*. New York: Doubleday.

Stanko, E. (1985). *Intimate intrusions: Women's experiences of male violence*. London: Routledge.

Starhawk. (1979). *The spiral dance: A rebirth of the ancient religion of the great goddess*. San Francisco, CA: HarperCollins.

Steger, M., & Roy, R. (2010). *Neoliberalism: A very short introduction*. New York: Oxford Press.

Stout, K. (1992). Intimate femicide: An ecological analysis. *Journal of Sociology and Social Welfare*, 19, 29–50.

Straus, M.A. (1991). Discipline and deviance: Physical punishment of children and violence and other crime in adulthood. *Social Problems*, 38, 133–154.

Straus, M.A. (1994). *Beating the devil out of them: Corporal punishment in American families*. New York: Lexington Books.

Stringer, R. (2014). *Knowing victims: Feminism, agency and victim politics in neoliberal times*. Hoboken, NJ: Taylor and Francis.

Stull, D.D., & Broadway, M.J. (2003). *Slaughterhouse blues: The meat and poultry industry in North America*. Belmont, CA: Thomason/Wadsworth.

Sturgeon, N. (1997). The nature of race: Discourses of racial difference in ecofeminism. In K.J. Warren (Ed.), *Ecofeminism: Women, Culture, Nature* (pp. 260–278). Bloomington: Indiana University Press.

Sturgeon, N. (2017). Facing the future, honouring the past: Whose gender? Whose nature? In S. MacGregor (Ed.). *Routledge handbook of gender and environment* (Routledge international handbooks) (pp. xxi–xxii). Abingdon, Oxon: Routledge.

Susskind, Y. (2018). Indigenous women's anti-violence strategies. In S. Bahun-Radunovic (Ed.), *Violence and gender in the globalized world: The intimate and the extimate* (pp. 9–31). Milton: Taylor and Francis.

Taylor, D., Hartmann, D., Dezecache, G., Te Wong, S., & Davila-Ross, M. (2019). Facial complexity in sun bears: Exact facial mimicry and social sensitivity. *Scientific Reports*, 9(1), 1–6. doi:10.1038/s41598-019-39932-6

Taylor, S. (2017). *Beasts of burden: Animal and disability liberation*. New York: The New Press.

Thompson, C., & MacGregor, S. (2017). The death of nature: Foundations of ecological feminist thought. In S. MacGregor (Ed.), *Routledge handbook of gender and environment* (Routledge international handbooks) (pp. 43–53). Abingdon, Oxon: Routledge.

Throsby, K., & Alexander, F. (Eds.). (2008). *Gender and interpersonal violence: Language, action and representation*. New York: Palgrave Macmillan.

Tickner, J.A. (2001). *Gendering world politics: Issues and approaches in the post-cold war era*. New York: Columbia University Press.

Titterington, V.A. (2006). A retrospective investigation of gender inequality and female homicide victimization. *Sociological Spectrum*, 26, 205–236.

Tjaden, P., & Thoennes, N. (2000). Full report of the prevalence, incidence, and consequences of violence against women. Research Report. National Institute of Justice and the Centers for Disease Control and Prevention: Atlanta, GA.

Torres, B. (2007). *Making a killing: The political economy of animal rights*. Edinburgh: AK Press.

Townsend, E.A. (2018). The Palgrave handbook of practical animal ethics. In *Suffering of animals in food production: Problems and practical solutions* (pp. 445–473). London: Palgrave Macmillan. doi:10.1057/978-1-137–36671-9_26

Tripp, A., Ferree, M., & Ewig, C. (Eds.). (2013). *Gender, violence, and human security: Critical feminist perspectives*. New York: New York University Press.

True, J. (2012). *The political economy of violence against women* (Oxford studies in gender and international relations). New York: Oxford University Press.

Turner, J.H. (1998). *The structure of sociological theory* (6th ed.). Belmont, CA: Wadsworth.

Turpin, J., & Lorentzen, L. (1996). *The gendered new world order: Militarism, development, and the environment*. New York: Routledge.

Twine, R. (2010). Intersectional disgust? Animals and (eco)feminism. *Feminism and Psychology*, 20(3), 397–406.

Twine, R. (2014). Ecofeminism and veganism: Revisiting the questions of universalism. In C.J. Adams & L. Gruen (Eds.), *Ecofeminism: Feminist intersections with other animals and the earth* (pp. 44–68). New York: Bloomsbury.

Uhlig, R. (2002). 10 million animals were slaughtered in foot and mouth cull. Retrieved November 22, 2018. *The Daily Telegraph, 09*, 09-09.

UNHCR (United Nations High Commissioner for Refugees). (2003). Sexual and gender-based Violence. Retrieved November, 2018, from https://www.unhcr.org/en-us/protection/women/3f696bcc4/sexual-gender-based-violence-against-refugees-returnees-internally-displaced.html

United States Department of Justice. (1996). *Female victims of violent crime* (Office of Justice Programs: Bureau of Justice Statistics, NCJ-162602). Retrieved September, 2018, https://static.prisonpolicy.org/scans/bjs/fvvc.pdfe.

United States Government Accountability Office, & Yager, L. (2011). *The democratic republic of the Congo: Information on the rate of sexual violence in war-torn eastern DRC and adjoining countries: Report to congressional committees*. Washington, DC: U.S. Govt. Accountability Office.

United Way of Santa Cruz County. (1990). *A post-earthquake community needs assessment for Santa Cruz County*. Aptos, CA: United Way of Santa Cruz County, 201.

Vance, L. (1997). Ecofeminism and wilderness. *NWSA Journal: A Publication of the National Women's Studies Association*, 9(3), 60–60.

Van Cleve, J. (1993). *Deaf history unveiled: Interpretations from the new scholarship*. Washington, DC: Gallaudet University Press.

Verchick, R.M. (2004). Feminist theory and environmental justice. In R. Stein (Ed.), *New perspectives on environmental justice: Gender, sexuality, and activism* (pp. 91–123). New Brunswick, NJ: Rutgers University Press.

Vieraitis, L.M., & Williams, M.R. (2002). Assessing the impact of gender inequality of female homicide victimization across U.S. cities: A racially disaggregated analysis. *Violence Against Women*, 8, 35–63.

Vieraitis, L.M., Britto, S., & Kovandzic, T.V. (2007). The impact of women's status and gender inequality on female homicide victimization rates: Evidence from U.S. counties. *Feminist Criminology*, 2, 57–73.

Villalón, R. (2010). *Violence against Latina immigrants: Citizenship, inequality, and community*. New York: New York University Press.

Waal, F. (1995). Bonobo sex and society. *Scientific American*, 272(3), 82–88.

Waal, F. (2016). *Are we smart enough to know how smart animals are?* (1st ed.). New York: W.W. Norton.

Waal, F. (2019). *Mama's last hug: Animal emotions and what they tell us about ourselves* (1st ed.). New York: W.W. Norton.

Walby, S. (1990). *Theorizing patriarchy*. Cambridge, MA: Basil Blackwell.

Walker, A. (1988). Am I blue? Ain't these tears in these eyes tellin' you? In I. Zahava (Ed.), *Through other eyes: Animal stories by women* (pp. 1–6). Freedom, CA: The Crossing Press.

Walker, L.E. (1977–1978). Battered women and learned helplessness. *Victimology*, 3–4, 525–534.

Walters, M.L. (2011). Straighten up and act like a lady: A qualitative study of lesbian survivors of intimate partner violence. *Journal of Gay & Lesbian Social Services*, 23(2), 250–270. doi:10.1080/10538720.2011.559148

Warkentin, T. (2012). Must every animal studies scholar be vegan? *Hypatia*, 27(3), 499–504.

Warren, K. (1987). Feminism and ecology: Making connections. *Environmental Ethics*, 9(1), 3.

Warren, K. (1990). The power and the promise of ecological feminism. *Environmental Ethics*, 12(2), 125.

Warren, K. (1996). *Ecological feminist philosophies*. Bloomington: Indiana University Press.

Warren, K., & Cady, D. (1994). Feminism and peace: Seeing connections. *Hypatia*, 9(2), 4–4.

Warren, K., & Cady, D. (1996). *Bringing peace home: Feminism, violence, and nature* (A Hypatia book). Bloomington: Indiana University Press.

Warren, K., & Wells-Howe, B. (1994). *Ecological feminism*. London: Routledge.

Warren, K.J. (2000). *Ecofeminist philosophy: A western perspective on what it is and why it matters*. Lanham, MD: Rowman & Littlefield.

Watson, S. (1991). Femocratic feminisms. *The Sociological Review*, 39, 186–204.

Whaley, R.B. (2001). The paradoxical relationship between gender inequality and rape: Toward a refined theory. *Gender and Society*, 15, 531–555.

Whaley, R.B., & Messner, S.F (2002). Gender equality and gendered homicides. *Homicide Studies*, 6, 188–210.

White, E. (1995). Black Women and Wilderness. In T. Jordan & J. Hepworth (Eds.), *The stories that shape us: Contemporary women write about the west: An anthology* (1st ed., pp. 13–33). New York: W.W. Norton.

White, R., & Heckenberg, D. (2014). *Green criminology: An introduction to the study of environmental harm* (1st ed.). New York: Routledge, Taylor & Francis Group.

Widom, C.S. (2000). Childhood victimization: Early adversity, later psychopathology. *National Institute of Justice Journal* [NCJ 180077] January, 2–9.

Williams, C. (2010). *Ecology and socialism: Solutions to capitalist ecological crisis*. Chicago, IL: Haymarket Books.

Wilson, J., Phillips, B.D., & Neal, D.M. (1998). Domestic violence after disaster. In E. Enarson & B.H. Morrow (Eds.), *The gendered terrain of disaster: Through women's eyes* (pp. 225–231). Westport, CT: Praeger.

Wolfe, C. (2003). *Animal rites: American culture, the discourse of species, and posthumanist theory*. Chicago, IL: University of Chicago Press.

Wolfe, C. (2010). *What is posthumanism?* Minneapolis: University of Minnesota Press.

Wolfe, C. (2013). *Before the law: Human and other animals in biopolitical frame*. Chicago, IL: Chicago University Press.

Woolf, V. (1938). *Three guineas* (1st American ed.). New York: Harcourt, Brace.

Wriggins, J. (1983). Rape, racism, and the law. *Harvard Women's Law Journal, 6*, 103–141.

Yllo, K.A. (1983). Sexual equality and violence against wives in American states. *Journal of Comparative Family Studies, 14*, 67–86.

Yllo, K.A. (1993). Through a feminist lens: Gender, power and violence. In R.J. Gelles & D.R. Loseke (Eds.), *Current controversies on family violence* (pp. 88–101). Newberry Park, CA: Sage.

Yllo, K.A., & Straus, M.A. (1984). The impact of structural inequality and sexist family norms on rates of wife beating. *Journal of International and Comparative Social Welfare, 1*, 16–29.

Young, I. (1990). *Justice and the politics of difference* (Princeton paperbacks). Princeton, NJ: Princeton University Press.

Zahn-Waxler, C., & Radke-Yarrow M (1990).The origins of empathic concern. *Motivation and Emotion, 14*(2), 107–130.

Zimbardo, P. (2009). *The Lucifer effect: How good people turn evil* (Pbk. ed.). London: Rider.

Zimmerman, M. (1994). *Contesting earth's future: Radical ecology and postmodernity*. Berkeley: University of California Press.

Žižek, S. (2008). *Violence*. New York: Picador.

Index

Printed and bound by CPI Group (UK) Ltd, Croydon, CR0 4YY

24/10/2024

01778282-0018